Introduction to IP and ATM
Design and Performance

Introduction to IP and ATM Design and Performance

With Applications Analysis Software

Second Edition

J M Pitts
J A Schormans

Queen Mary
University of London
UK

JOHN WILEY & SONS, LTD
Chichester • New York • Weinheim • Brisbane • Toronto • Singapore

First Edition published in 1996 as *Introduction to ATM Design and Performance* by John Wiley & Sons, Ltd.
Copyright © 2000 by John Wiley & Sons, Ltd

Baffins Lane, Chichester,
West Sussex, PO19 1UD, England

National   01243 779777
International (+44) 1243 779777
e-mail (for orders and customer service enquiries): cs-books@wiley.co.uk

Visit our Home Page on http://www.wiley.co.uk or http://www.wiley.com

Other Wiley Editorial Offices

John Wiley & Sons, Inc., 605 Third Avenue,
New York, NY 10158-0012, USA

Wiley-VCH Verlag GmbH
Pappelallee 3, D-69469 Weinheim, Germany

Jacaranda Wiley Ltd, 33 Park Road, Milton,
Queensland 4064, Australia

John Wiley & Sons (Canada) Ltd, 22 Worcester Road
Rexdale, Ontario, M9W 1L1, Canada

John Wiley & Sons (Asia) Pte Ltd, 2 Clementi Loop #02-01,
Jin Xing Distripark, Singapore 129809

British Library Cataloguing in Publication Data

A catalogue record for this book is available from the British Library

ISBN 0471 49187 X

Typeset in $10\frac{1}{2}/12\frac{1}{2}$pt Palatino by Laser Words, Chennai, India
Printed and bound in Great Britain by Bookcraft (Bath) Ltd
This book is printed on acid-free paper responsibly manufactured from sustainable forestry, in which at least two trees are planted for each one used for paper production.

To

Suzanne, Rebekah, Verity and Barnabas

Jacqueline, Matthew and Daniel

Contents

Preface   xi

PART I   INTRODUCTORY TOPICS   1

1   An Introduction to the Technologies of IP and ATM   3

Circuit Switching   3
Packet Switching   5
Cell Switching and ATM   7
Connection-orientated Service   8
Connectionless Service and IP   9
Buffering in ATM switches and IP routers   11
Buffer Management   11
Traffic Control   13

2   Traffic Issues and Solutions   15

Delay and Loss Performance   15
    Source models   16
    Queueing behaviour   18
Coping with Multi-service Requirements: Differentiated Performance   30
    Buffer sharing and partitioning   30
    Cell and packet discard mechanisms   32
    Queue scheduling mechanisms   35
Flows, Connections and Aggregates   37
    Admission control mechanisms   37
    Policing mechanisms   40
    Dimensioning and configuration   41

3   Teletraffic Engineering   45

Sharing Resources   45
Mesh and Star Networks   45
Traffic Intensity   47
Performance   49
TCP: Traffic, Capacity and Performance   49
Variation of Traffic Intensity   50
Erlang's Lost Call Formula   52
Traffic Tables   53

4 Performance Evaluation   57

Methods of Performance Evaluation   57
   Measurement   57
   Predictive evaluation: analysis/simulation   57
Queueing Theory   58
   Notation   60
   Elementary relationships   60
   The M/M/1 queue   61
   The M/D/1/K queue   64
   Delay in the M/M/1 and M/D/1 queueing systems   65

5 Fundamentals of Simulation   69

Discrete Time Simulation   69
   Generating random numbers   71
   M/D/1 queue simulator in Mathcad   73
   Reaching steady state   74
   Batch means and confidence intervals   75
   Validation   77
Accelerated Simulation   77
   Cell-rate simulation   77

6 Traffic Models   81

Levels of Traffic Behaviour   81
Timing Information in Source Models   82
Time between Arrivals   83
Counting Arrivals   86
Rates of Flow   89

PART II   ATM QUEUEING AND TRAFFIC CONTROL   95

7 Basic Cell Switching   97

The Queueing Behaviour of ATM Cells in Output Buffers   97
Balance Equations for Buffering   98
Calculating the State Probability Distribution   100
Exact Analysis for FINITE Output Buffers   104
Delays   108
   End-to-end delay   110

8 Cell-Scale Queueing   113

Cell-scale Queueing   113
Multiplexing Constant-bit-rate Traffic   114
Analysis of an Infinite Queue with Multiplexed CBR Input: The   115
N·D/D/1
Heavy-traffic Approximation for the M/D/1 Queue   117
Heavy-traffic Approximation for the N·D/D/1 Queue   119
Cell-scale Queueing in Switches   121

9 Burst-Scale Queueing   125

ATM Queueing Behaviour   125
Burst-scale Queueing Behaviour   127

Fluid-flow Analysis of a Single Source – Per-VC Queueing 129
Continuous Fluid-flow Approach 129
Discrete 'Fluid-flow' Approach 131
Comparing the Discrete and Continuous Fluid-flow Approaches 136
Multiple ON/OFF Sources of the Same Type 139
The Bufferless Approach 141
The Burst-scale Delay Model 145

10 Connection Admission Control 149
The Traffic Contract 150
Admissible Load: The Cell-scale Constraint 151
A CAC algorithm based on M/D/1 analysis 152
A CAC algorithm based on N·D/D/1 analysis 153
The cell-scale constraint in statistical-bit-rate transfer capability, based on 155
M/D/1 analysis
Admissible Load: The Burst Scale 157
A practical CAC scheme 159
Equivalent cell rate and linear CAC 160
Two-level CAC 160
Accounting for the burst-scale delay factor 161
CAC in The Standards 165

11 Usage Parameter Control 167
Protecting the Network 167
Controlling the Mean Cell Rate 168
Algorithms for UPC 172
The leaky bucket 172
Peak Cell Rate Control using the Leaky Bucket 173
The problem of tolerances 176
Resources required for a worst-case ON/OFF cell stream from peak cell 178
rate UPC
Traffic shaping 182
Dual Leaky Buckets: The Leaky Cup and Saucer 182
Resources required for a worst-case ON/OFF cell stream from sustainable 184
cell rate UPC

12 Dimensioning 187
Combining The Burst and Cell Scales 187
Dimensioning The Buffer 190
Small buffers for cell-scale queueing 193
Large buffers for burst-scale queueing 198
Combining The Connection, Burst and Cell Scales 200

13 Priority Control 205
Priorities 205
Space Priority and The Cell Loss Priority Bit 205
Partial Buffer Sharing 207
Increasing the admissible load 214
Dimensioning buffers for partial buffer sharing 215
Time Priority in ATM 218
Mean value analysis 219

PART III   IP PERFORMANCE   **227**
**   AND TRAFFIC MANAGEMENT**

14   Basic Packet Queueing   **229**

The Queueing Behaviour of Packets in an IP Router Buffer   229
Balance Equations for Packet Buffering: The Geo/Geo/1   230
    Calculating the state probability distribution   231
Decay Rate Analysis   234
    Using the decay rate to approximate the buffer overflow probability   236
Balance Equations for Packet Buffering: Excess-rate Queueing   238
Analysis
    The excess-rate M/D/1, for application to voice-over-IP   239
    The excess-rate solution for best-effort traffic   245

15   Resource Reservation   **253**

Quality of Service and Traffic Aggregation   253
Characterizing an Aggregate of Packet Flows   254
Performance Analysis of Aggregate Packet Flows   255
    Parameterizing the two-state aggregate process   257
    Analysing the queueing behaviour   259
Voice-over-IP, Revisited   261
Traffic Conditioning of Aggregate Flows   265

16   IP Buffer Management   **267**

First-in First-out Buffering   267
Random Early Detection – Probabilistic Packet Discard   267
Virtual Buffers and Scheduling Algorithms   273
    Precedence queueing   273
    Weighted fair queueing   274
Buffer Space Partitioning   275
Shared Buffer Analysis   279

17   Self-similar Traffic   **287**

Self-similarity and Long-range-dependent Traffic   287
The Pareto Model of Activity   289
Impact of LRD Traffic on Queueing Behaviour   292
    The Geo/Pareto/1 Queue   293

References   **299**

Index   **301**

Preface

In recent years, we have taught design and performance evaluation techniques to undergraduates and postgraduates in the Department of Electronic Engineering at Queen Mary and Westfield College (http://www.elec.qmw.ac.uk/) and to graduates on various University of London M.Sc. courses for industry. We have found that many engineers and students of engineering experience difficulty in making sense of teletraffic issues. This is partly because of the subject itself: the technologies and standards are flexible, complicated, and always evolving. However, some of the difficulties arise because of the advanced mathematical models that have been applied to IP and ATM analysis. The research literature, and many books reporting on it, is full of differing analytical approaches applied to a bewildering array of traffic mixes, buffer management mechanisms, switch designs, and traffic and congestion control algorithms.

To counter this trend, our book, which is intended for use by students both at final-year undergraduate, and at postgraduate level, and by practising engineers in the telecommunications and Internet world, provides an *introduction* to the design and performance issues surrounding IP and ATM. We cover performance evaluation by analysis and simulation, presenting key formulas describing traffic and queueing behaviour, and practical examples, with graphs and tables for the design of IP and ATM networks.

In line with our general approach, derivations are included where they demonstrate an intuitively simple technique; alternatively we give the formula (and a reference) and then show how to apply it. As a bonus, the formulas are available as Mathcad files (see below for details) so there is no need to program them for yourself. In fact, many of the graphs have the Mathcad code right beside them on the page. We have ensured that the need for prior knowledge (in particular, probability theory) has been kept to a minimum. We feel strongly that this enhances the work, both as a textbook and as a design guide; it is far easier to make progress when you are not trying to deal with another subject in the background.

For the second edition, we have added a substantial amount of new material on IP traffic issues. Since the first edition, much work has been done in the IP community to make the technology QoS-aware. In essence, the techniques and mechanisms to do this are generic – however, they are often disguised by the use of confusing jargon in the different communities. Of course, there are real differences in the technologies, but the underlying approaches for providing guaranteed performance to a wide range of service types are very similar.

We have introduced new ideas from our own research – more accurate, usable results and understandable derivations. These new ideas make use of the excess-rate technique for queueing analysis, which we have found applicable to a wide variety of queueing systems. Whilst we still do not claim that the book is comprehensive, we do believe it presents the essentials of design and performance analysis for both IP and ATM technologies in an intuitive and understandable way.

Applications analysis software

Where's the disk or CD? Unlike the first edition, we decided to put all the Mathcad files on a web-site for the book. But in case you can't immediately reach out and click on the Internet, most of the figures in the book have the Mathcad code used to generate them alongside, so take a look. Note that where Mathcad functions have been defined for previous figures, they are not repeated, for clarity. So, check out http://www.elec.qmw.ac.uk/ipatm. You'll also find some homework problems there.

Organization

In Chapter 1, we describe both IP and ATM technologies. On the surface the technologies appear to be rather different, but both depend on similar approaches to buffer management and traffic control in order to provide performance guarantees to a wide variety of services. We highlight the fundamental operations of both IP and ATM as they relate to the underlying queueing and performance issues, rather than describe the technologies and standards in detail. Chapter 2 is the executive summary for the book: it gathers together the range of analytical solutions covered, lists the parameters, and groups them according to their use in addressing IP and ATM traffic issues. You may wish to skip over it on a first reading, but use it afterwards as a ready reference.

Chapter 3 introduces the concept of resource sharing, which underpins the design and performance of any telecommunications technology, in the context of circuit-switched networks. Here, we see the trade-off between the economics of providing telecommunications capability and satisfying the service requirements of the customer. To evaluate the performance of shared resources, we need an understanding of queueing theory. In Chapter 4, we introduce the fundamental concept of a queue (or waiting line), its notation, and some elementary relationships, and apply these to the basic process of buffering, using ATM as an example. This familiarizes the reader with the important measures of delay and loss (whether of packets or cells), the typical orders of magnitude for these measures, and the use of approximations, *without* having to struggle through analytical derivations at the same time. Simulation is widely used to study performance and design issues, and Chapter 5 provides an introduction to the basic principles, including accelerated techniques.

Chapter 6 describes a variety of simple traffic models, both for single sources and for aggregate traffic, with sample parameter values typical of IP and ATM. The distinction between levels of traffic behaviour, particularly the cell/packet and burst levels is introduced, as well as the different ways in which timing information is presented in source models. Both these aspects are important in helping to simplify and clarify the analysis of queueing behaviour.

In Part II, we turn to queueing and traffic control issues, with the specific focus on ATM. Even if your main interest is in IP, we recommend you read these chapters. The reason is not just that the queueing behaviour is very similar (ATM cells and fixed-size packets look the same to a queueing system), but because the development of an appreciation for both the underlying queueing issues and the influence of key traffic parameters builds in a more intuitive way.

In Chapter 7, we treat the queueing behaviour of ATM cells in output buffers, taking the reader very carefully through the analytical derivation of the queue state probability distribution, the cell loss probability, and the cell delay distribution. The analytical approach used is a direct probabilistic technique which is simple and intuitive, and key stages in the derivation are illustrated graphically. This basic technique is the underlying analytical approach applied in Chapter 13 to the more complex issues of priority mechanisms, in Chapter 14 to basic packet switching with variable-length packets, and in Chapter 17 to the problem of queueing under self-similar traffic input.

Chapters 8 and 9 take the traffic models of Chapter 6 and the concept of different levels of traffic behaviour, and apply them to the analysis of ATM queueing. The distinction between cell-scale queueing (Chapter 8) and burst-scale queueing (Chapter 9) is of fundamental importance because it provides the basis for understanding and designing a traffic control framework (based on the international standards for ATM) that can handle integrated, multi-service traffic mixes. This framework is described in Chapters 10, 11, 12 and 13. A key part of the treatment of cell- and burst-scale queueing is the use of explicit formulas, based on heavy-traffic approximate analysis; these formulas can be rearranged very simply to illustrate the design of algorithms for connection admission control (Chapter 10), usage parameter control (Chapter 11), and buffer dimensioning (Chapter 12). In addition, Chapter 12 combines the cell- and burst-scale analysis with the connection level for link dimensioning, by incorporating Erlang's loss analysis introduced in Chapter 3. In Chapter 13, we build on the analytical approach, introduced in Chapter 7, to cover space and time priority issues.

Part III deals with IP and its performance and traffic management. Chapter 14 applies the simple queueing analysis from Chapter 7 to the buffering of variable-size packets. A new approximate technique, based on the notion of excess-rate, is developed for application to queueing systems with fixed, bi-modal, or general packet size distributions. The technique gives accurate results across the full range of load values, and has wide applicability in both IP and ATM. The concept of decay rate is introduced; decay rate is a very flexible tool for summarising queueing behaviour, and is used to advantage in the following chapters. Chapter 15 addresses the issue of resource reservation for aggregate flows in IP. A full burst-scale analysis, applicable to both delay-sensitive, and loss-sensitive traffic, is developed by judicious parameterization of a simple two-state model for aggregate packet flows. The analysis is used to produce design curves for configuring token buckets: for traffic conditioning of behaviour aggregates in Differentiated Services, or for queue scheduling of traffic flows in the Integrated Services Architectures.

Chapter 16 addresses the topic of buffer management from an IP perspective, relying heavily on the use of decay rate analysis from previous chapters. Decay rates are used to illustrate the configuration of thresholds in the probabilistic packet discard mechanism known as RED (random early detection). The partitioning of buffer space and service capacity into virtual buffers is introduced, and simple techniques for configuring buffer partitions, based on decay rate analysis, are developed.

Finally, in Chapter 17, we give a simple introduction to the important, and mathematically challenging, subjects of self-similarity and long-range dependence. We illustrate these issues with the Pareto distribution as a traffic model, and show its impact on queueing behaviour using the simple analysis developed in Chapter 7.

Acknowledgements

This new edition has benefited from the comments and questions raised by readers of the first edition, posted, e-mailed and telephoned from around the world. We would like to thank our colleagues in the Department of Electronic Engineering for a friendly, encouraging and stimulating academic environment in which to work. But most important of all are our families – thank you for your patience, understanding and support through thick and thin!

PART I

Introductory Topics

1 An Introduction to the Technologies of IP and ATM

the bare necessities

This chapter is intended as a brief introduction to the technologies of the Asynchronous Transfer Mode (ATM) and the Internet Protocol (IP) on the assumption that you will need some background information before proceeding to the chapters on traffic engineering and design. If you already have a good working knowledge you may wish to skip this chapter, because we highlight the fundamental operation *as it relates to performance issues* rather than describe the technologies and standards in detail. For anyone wanting a deeper insight we refer to [1.1] for a comprehensive introduction to the narrowband Integrated Services Digital Network (ISDN), to [1.2] for a general introduction to ATM (including its implications for interworking and evolution) and to [1.3] for next-generation IP.

CIRCUIT SWITCHING

In traditional analogue circuit switching, a call is set-up on the basis that it receives a path (from source to destination) that is its 'property' for the duration of the call, i.e. the whole of the bandwidth of the circuit is available to the calling parties for the whole of the call. In a digital circuit-switched system, the whole bit-rate of the line is assigned to a call for only a single time slot per frame. This is called 'time division multiplexing'.

During the time period of a frame, the transmitting party will generate a fixed number of bits of digital data (for example, 8 bits to represent the level of an analogue telephony signal) and these bits will be grouped together in the time slot allocated to that call. On a transmission link, the same time slot in every frame is assigned to a call for the duration of that call (Figure 1.1). So the time slot is identified by its *position* in the frame, hence use of the name 'position multiplexing', although this term is not used as much as 'time division multiplexing'.

When a connection is set up, a route is found through the network and that route remains fixed for the duration of the connection. The route will probably traverse a number of switching nodes and require the use of many transmission links to provide a circuit from source to destination. The time slot position used by a call is likely to be different on each link. The switches which interconnect the transmission links perform the time slot interchange (as well as the space switching) necessary to provide the 'through-connection' (e.g. link M, time slot 2 switches to link N, time slot 7 in Figure 1.2).

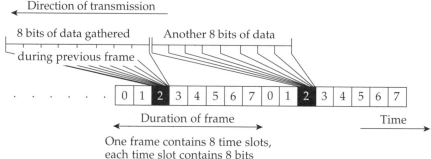

Figure 1.1.   An Example of Time Division, or Position, Multiplexing

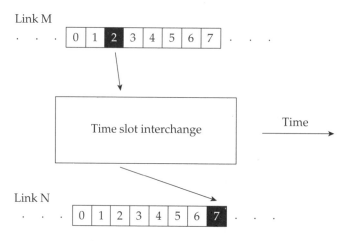

Figure 1.2.   Time Slot Interchange

In digital circuit-switched telephony networks, frames have a repetition rate of 8000 frames per second (and so a duration of 125 µs), and as there are always 8 bits (one byte) per time slot, each channel has a bit-rate of 64 kbit/s. With N time slots in each frame, the bit-rate of the line is $N \cdot 64$ kbit/s. In practice, extra time slots or bits are added for control and synchronization functions. So, for example, the widely used 30-channel system has two extra time slots, giving a total of 32 time slots, and thus a bit-rate of $(30 + 2) \times 64 = 2048$ kbit/s. Some readers may be more familiar with the 1544 kbit/s 24-channel system which has 1 extra bit per frame.

The time division multiplexing concept can be applied recursively by considering a 24- or 30-channel system as a single 'channel', each frame of which occupies one time slot per frame of a higher-order multiplexing system. This is the underlying principle in the synchronous digital hierarchy (SDH), and an introduction to SDH can be found in [1.1].

The main performance issue for the user of a circuit-switched network is whether, when a call is requested, there is a circuit available to the required destination. Once a circuit is established, the user has available a constant bit-rate with a fixed end-to-end delay. There is no error detection or correction provided by the network on the circuit – that's the responsibility of the terminals at either end, if it is required. Nor is there any per circuit overhead – the whole bit-rate of the circuit is available for user information.

PACKET SWITCHING

Let's now consider a generic packet-switching network, i.e. one intended to represent the main characteristics of packet switching, rather than any particular packet-switching system (later on in the chapter we'll look more closely at the specifics of IP).

Instead of being organized into single eight-bit time slots which repeat at regular intervals, data in a packet-switched network is organised into packets comprising *many* bytes of user data (bytes may also be known as 'octets'). Packets can vary in size depending on how much data there is to send, usually up to some predetermined limit (for example, 4096 bytes). Each packet is then sent from node to node as a group of contiguous bits fully occupying the link bit-rate for the duration of the packet. If there is no packet to send, then nothing is sent on the link. When a packet is ready, and the link is idle, then the packet can be sent immediately. If the link is busy (another packet is currently being transmitted), then the packet must wait in a buffer until the previous one has completed transmission (Figure 1.3).

Each packet has a label to identify it as belonging to a particular communication. Thus packets from different sources and to different

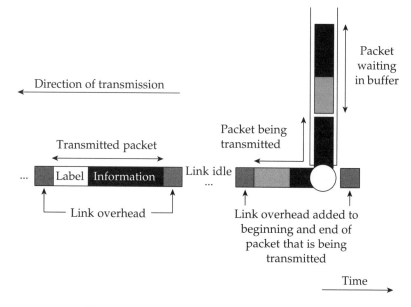

Figure 1.3.   An Example of Label Multiplexing

destinations can be multiplexed over the same link by being transmitted one after the other. This is called 'label multiplexing'. The label is used at each node to select an outgoing link, routeing the packet across the network. The outgoing link selected may be predetermined at the set-up of the connection, or it may be varied according to traffic conditions (e.g. take the least busy route). The former method ensures that packets arrive in the order in which they were sent, whereas the latter method requires the destination to be able to resequence out-of-order packets (in the event that the delays on alternative routes are different).

Whichever routeing method is used, the packets destined for a particular link must be queued in the node prior to transmission. It is this queueing which introduces variable delay to the packets. A system of acknowledgements ensures that errored packets are not lost but are retransmitted. This is done on a link-by-link basis, rather than end-to-end, and contributes further to the variation in delay if a packet is corrupted and needs retransmission. There is quite a significant per-packet overhead required for the error control and acknowledgement mechanisms, in addition to the label. This overhead reduces the effective bit-rate available for the transfer of user information. The packet-plus-link overhead is often (confusingly) called a 'frame'. Note that it is *not* the same as a frame in circuit switching.

A simple packet-switched network may continue to accept packets without assessing whether it can cope with the extra traffic or not. Thus it appears to be non-blocking, in contrast to a circuit-switched network which rejects (blocks) a connection request if there is no circuit available. The effect of this non-blocking operation is that packets experience greater and greater delays across the network, as the load on the network increases. As the load approaches the network capacity, the node buffers become full, and further incoming packets cannot be stored. This triggers retransmission of those packets which only worsens the situation by increasing the load; the successful throughput of packets decreases significantly.

In order to maintain throughput, congestion control techniques, particularly flow control, are used. Their aim is to limit the rate at which sources offer packets to the network. The flow control can be exercised on a link-by-link, or end-to-end basis. Thus a connection cannot be *guaranteed* any particular bit-rate: it is allowed to send packets to the network as and when it needs to, but if the network is congested then the network exerts control by restricting this rate of flow.

The main performance issues for a user of a packet-switched network are the delay experienced on any connection and the throughput. The network operator aims to maximize throughput and limit the delay, even in the presence of congestion. The user is able to send information on demand, and the network provides error control through re-transmission of packets on a link-by-link basis. Capacity is not dedicated to the connection, but shared on a dynamic basis with other connections. The capacity available to the user is reduced by the per-packet overheads required for label multiplexing, flow and error control.

CELL SWITCHING AND ATM

Cell switching combines aspects of both circuit and packet switching. In very simple terms, the ATM concept maintains the time-slotted nature of transmission in circuit switching (but without the position in a frame having any meaning) but increases the size of the data unit from one octet (byte) to 53 octets. Alternatively, you could say that ATM maintains the concept of a packet but restricts it to a fixed size of 53 octets, and requires packet-synchronized transmission.

This group of 53 octets is called a 'cell'. It contains 48 octets for user data – the information field – and 5 octets of overhead – the header. The header contains a label to identify it as belonging to a particular connection. So ATM uses label multiplexing and not position multiplexing. But what about the time slots? Well, these are called 'cell slots'. An ATM link operates a sort of conveyor belt of cell slots (Figure 1.4). If there is a cell to send, then it must wait for the start of the next cell slot boundary – the next slot on the conveyor belt. The cell is not allowed to straddle two slots. If there is no cell to send, then the cell slot is unused, i.e. it is empty.

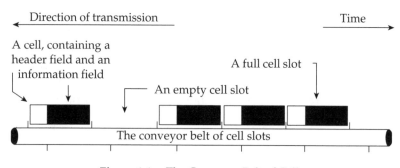

Figure 1.4.   The Conveyor Belt of Cells

There is no need for the concept of a repeating frame, as in circuit switching, because the label in the header identifies the cell.

CONNECTION-ORIENTATED SERVICE

Let's take a more detailed look at the cell header in ATM. The label consists of two components: the virtual channel identifier (VCI) and the virtual path identifier (VPI). These identifiers do not have end-to-end (user-to-user) significance; they identify a particular virtual channel (VC) or virtual path (VP) on the link over which the cell is being transmitted. When the cell arrives at the next node, the VCI and the VPI are used to look up in the routeing table to what outgoing port the cell should be switched and what new VCI and VPI values the cell should have. The routeing table values are established at the set-up of a connection, and remain constant for the duration of the connection, so the cells always take the same route through the network, and the 'cell sequence integrity' of the connection is maintained. Hence ATM provides *connection-orientated* service.

But surely only one label is needed to achieve this cell routeing mechanism, and that would also make the routeing tables simpler: so why have two types of identifier? The reason is for the flexibility gained in handling connections. The basic equivalent to a circuit-switched or packet-switched connection in ATM is the virtual channel connection (VCC). This is established over a *series* of concatenated virtual channel links. A virtual path is a bundle of virtual channel links, i.e. it groups a number of VC links *in parallel*. This idea enables direct 'logical' routes to be established between two switching nodes that are not connected by a direct physical link.

The best way to appreciate why this concept is so flexible is to consider an example. Figure 1.5 shows three switching nodes connected in a physical star structure to a 'cross-connect' node. Over this physical network, a logical network of three virtual paths has been established.

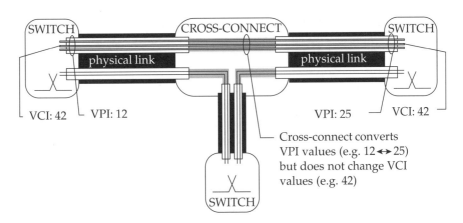

Figure 1.5. Virtual Paths and Virtual Channels

These VPs provide a logical mesh structure of virtual channel links between the switching nodes. The routeing table in the cross-connect only deals with port numbers and VPIs – the VCI values are neither read nor are they altered. However, the routeing table in the switching nodes deal with all three: port numbers, VPIs and VCIs.

In setting up a VCC, the cross-connect is effectively invisible; it does not need to know about VCIs and is therefore not involved in the process. If there was only one type of identifier in the ATM cell header, then either direct physical links would be needed between each pair of switching nodes to create a mesh network, or another switching node would be required at the hub of the star network. This hub switching node would then have to be involved in every connection set-up on the network.

Thus the VP concept brings significant benefits by enabling flexible logical network structures to be created to suit the needs of the expected traffic demands. It is also much simpler to change the logical network structure than the physical structure. This can be done to reflect, for example, time-of-day changes in demand to different destinations.

In some respects the VP/VC concept is rather similar to having a two-level time division multiplexing hierarchy in a circuit-switched network. It has extra advantages in that it is not bound by any particular framing structure, and so the capacity used by the VPs and VCs can be allocated in a very flexible manner.

CONNECTIONLESS SERVICE AND IP

So, ATM is a connection-orientated system: no user information can be sent across an ATM network unless a VCC or VPC has already been established. This has the advantage that real-time services, such as voice, do not have to suffer further delays associated with re-ordering the data on reception (in addition to the queueing and transmission delays experienced *en route*). However, for native connectionless-type services, the overhead of connection establishment, using signalling protocols, is burdensome.

Studies of Internet traffic have shown that the majority of flows are very short, comprising only a few packets, although the majority of packets belong to a minority of (longer-term) flows. Thus for the majority of flows, the overhead of signalling can exceed the amount of user information to be sent. IP deals with this in a very flexible way: it provides a *connectionless service* between end users in which successive data units can follow different paths. At a router, each packet is treated independently concerning the routeing decision (based on the destination address in the IP packet header) for the next hop towards the destination. This is ideal for transferring data on flows with small numbers of packets, and also works well for large numbers of packets. Thus packets being sent between the same source and destination points may follow quite different paths from source to destination.

Routes in IP are able to adapt quickly to congestion or equipment failure. Although from the point of view of each packet the service is in essence unreliable, for communication between end users IP is very robust. It is the transport-layer protocols, such as the transmission control protocol (TCP), that make up for the inherent unreliability of packet transfer in IP. TCP re-orders mis-sequenced packets, and detects and recovers from packet loss (or excessive delay) through a system of timers, acknowledgements and sequence numbers. It also provides a credit-based flow control mechanism which reacts to network congestion by reducing the rate at which packets are sent.

This is fine for elastic traffic, i.e. traffic such as email or file transfer that can adjust to large changes in delay and throughput (so long as the data eventually gets there), but not for stream, i.e. inelastic, traffic. This latter requires at least a minimum bit-rate across the network to be of any value. Voice telephony, at the normal rate of 64 kbit/s is an example wherein (unless some extra sophistication is present) this is the rate that must be supported otherwise the signal will suffer so much loss as to render it unintelligible and therefore meaningless. Requirements for inelastic traffic are difficult to meet, and impossible to guarantee, in an environment of highly variable delay, throughput and congestion. This is why they have traditionally been carried by technologies which are connection-orientated.

So, how can an IP network cope with both types of traffic, elastic and inelastic? The first requirement is to partition the traffic into groups that can be given different treatment appropriate to their performance needs. The second requirement is to provide the means to state their needs, and mechanisms to reserve resources specifically for those different groups of traffic. The Integrated Services Architecture (ISA), Resource Reservation Protocol (RSVP), Differentiated Services (DiffServ), and Multiprotocol Label Switching (MPLS) are a variety of means aimed at achieving just that.

BUFFERING IN ATM SWITCHES AND IP ROUTERS

Both IP and ATM networks move data about in discrete units. Network nodes, whether handling ATM cells or IP packets, have to merge traffic streams from different sources and forward them to different destinations via transmission links which the traffic shares for part of the journey. This process involves the temporary storage of data in finite-sized buffers, the actual pattern of arrivals causing queues to grow and diminish in size. Thus, in either technology, the data units contend for output transmission capacity, and in so doing form queues in buffers. In practice these buffers can be located in different places within the devices (e.g. at the inputs, outputs or crosspoints) but this is not of prime importance. The point is that queues form when the number of arrivals over a period exceeds the number of departures, and it is therefore the actual pattern of arrivals that is of most significance.

Buffering, then, is common to both technologies. However, simply providing buffers is not a good enough solution; it is necessary to provide the quality of service (QoS) that users request (and have to pay for). To ensure guaranteed QoS these buffers must be used intelligently, and this means providing *buffer management*.

BUFFER MANAGEMENT

Both ATM and IP feature buffer management mechanisms that are designed to enhance the capability of the networks. In essence, these mechanisms deal with how cells or packets gain access to the finite waiting area of the buffer and, once in that waiting area, how they gain access to the server for onward transmission. The former deals with how the buffer space is partitioned, and the discard policies in operation. The latter deals with how the packets or cells are ordered and scheduled for service, and how the service capacity is partitioned.

The key requirement is to provide partitions, i.e. *virtual* buffers, through which different groups of traffic can be forwarded. In the extreme, a virtual buffer is provided for each IP flow, or ATM connection, and it has its own buffer space and service capacity allocation. This is called per-flow or per-VC queueing. Typically, considerations of scale mean that traffic, whether flows or connections, must be handled in aggregate through virtual buffers, particularly in the core of the network. Terminology varies (e.g. transfer capability in ATM, behaviour aggregate in DiffServ, traffic trunk in MPLS), but the grouping tends to be according to traffic type, i.e. those with similar performance requirements and traffic characteristics.

Discard policies provide the means to differentiate between the relative magnitudes of data loss, and the extent and impact of loss on flows or connections within an aggregate of traffic.

In ATM, the cell loss priority (CLP) bit in the cell header is used to distinguish between two levels of what is called 'space priority'. A (virtual) buffer which is deemed full for low-priority cells can still allow high-priority cells to enter. The effect is to increase the likelihood of loss for low-priority cells compared with that for high-priority cells. Hence the name 'cell loss priority' bit. What use is this facility? Well, if a source is able to distinguish between information that is absolutely vital to its correct operation (e.g. video synchronization) and information which is not quite so important (e.g. part of the video picture) then the network can take advantage of the differences in cell loss requirement by accepting more traffic on the network. Otherwise the network loading is restricted by the more stringent cell loss requirement.

IP does have something similar: the type of service (ToS) field in IPv4 or the priority field in IPv6. Both fields have codes that specify different levels of loss treatment. In addition, in IP there is a discard mechanism called 'random early detection' (RED) that anticipates congestion by discarding packets *probabilistically* before the buffer becomes full. Packets are discarded with increasing probability when the average queue size is above a configurable threshold.

The rationale behind the RED mechanism derives from the particular challenge of forwarding best-effort packet traffic: TCP, in particular, can introduce unwelcome behaviour when the network (or part of it) is congested. If a buffer is full and has to discard arriving packets from many TCP connections, they will all enter their slow start phase. This reduces the load through the buffer, but because it affects so many connections it leads to a period of under-utilization. When all those TCP connections come out of slow start at about the same time, there is a substantial increase in traffic, causing congestion again in the buffer and more packet discard. The principle behind RED is that it applies the brakes gradually: in the early stages of congestion, only a few TCP connections are affected, and this may be sufficient to reduce the load and avoid any further increase in congestion. If the average queue size continues to increase, then packets are discarded with increasing probability, and so more TCP connections are affected. Once the average queue size exceeds an upper threshold all arriving packets are discarded.

In addition to controlling admission to the buffers, ATM and IP feature the ability to control the process that outputs data from the buffers – the buffer scheduling mechanism. This provides a means to differentiate between delay, as well as bit-rate, requirements: just as some traffic is more time-sensitive, so some needs greater transmission capacity. In both ATM and IP, the switches and routers can implement time priority ordering (precedence queueing) and mechanisms such as weighted fair queueing or round robin scheduling, to partition the service capacity among the virtual buffers.

TRAFFIC CONTROL

We have seen that both IP and ATM provide temporary storage for packets and cells in buffers across the network, introducing variable delays, and on occasion, loss too. These buffers incorporate various mechanisms to enable the networks to cater for different types of traffic – both elastic and inelastic. As we have noted, part of the solution to this problem is the use of buffer management strategies: partitioning and reserving appropriate resources – both buffer space and service capacity. However, there is another part to the overall solution: *traffic control*. This allows users to state their communications needs, and enables the network to coordinate and monitor its corresponding provision.

Upon receiving a reservation request (for a connection in ATM, or a flow in IP), a network assesses whether or not it can handle the traffic, in addition to what has already been accepted on the network. This process is rather more complicated than for circuit switching, because some of the reservation requests will be from variable bit-rate (VBR) services, for which the instantaneous bit-rate required will be varying in a random manner over time, as indeed will be the capacity available because many of the existing connections will also be VBR! So if a request arrives for a time-varying amount of capacity, and the capacity available is also varying with time, it is no longer a trivial problem to determine whether the connection or flow should be accepted.

In practice such a system works in the following way: the user declares values for some parameters which describe the traffic behaviour of the requested connection or flow, as well as the loss and delay performance required; the network then uses these traffic and performance values to come to an accept/reject decision, and informs the user. If accepted, the network has to ensure that the sequence of cells or packets corresponds to the declared traffic values. This whole process is aimed at *preventing* congestion in the network and ensuring that the performance requirements are met for *all* carried traffic.

The traffic and performance values agreed by the user and the network form a *traffic contract*. The mechanism which makes the accept/reject decision is the *admission control* function, and this resides in the ATM switching nodes or IP routers in the network. A mechanism is also necessary to ensure compliance with the traffic contract, i.e. the user should not exceed the peak (or mean, or whatever) rate that was agreed for the connection, flow, or aggregate. This mechanism is called *usage parameter control* (UPC) in ATM and is situated on entry to the network. If the user does exceed the traffic contract, then the UPC mechanism takes action to protect the network from the effects of this excess, e.g. discarding some of the cells from the non-compliant connection. A similar mechanism in DiffServ for IP networks is called *traffic conditioning*, and this involves packet metering, marking, shaping and dropping.

In order to design algorithms for these mechanisms, it is important that we understand the characteristics of the traffic sources, and the effects these sources have when they are multiplexed through buffers in the network, in terms of the delay and loss incurred. How we design the algorithms is very closely related to how large we make the buffers, and what buffer management mechanisms are proposed. Buffer dimensioning and management mechanisms depend on how we intend to handle the different services and their performance requirements.

2 Traffic Issues and Solutions

short circuits, short packets

This chapter is the executive summary for the book: it provides a quick way to find a range of analytical solutions for a variety of design and performance issues relating to IP and ATM traffic problems. If you are already familiar with performance evaluation and want a quick overview of what the book has to offer, then read on. Otherwise, you'll probably find that it's best to skip this chapter, and come back to it after you have read the rest of the book – you'll then be able to use this chapter as a ready reference.

DELAY AND LOSS PERFORMANCE

In cell- or packet-based networks, the fundamental behaviour affecting performance is the queueing experienced by cells/packets traversing the buffers within those switches or routers on the path(s) from source to destination through the network. This queueing behaviour means that cells/packets experience variations in the delay through a buffer and also, if that delay becomes too large, loss.

At its simplest, a buffer has a fixed service rate, a finite capacity for the temporary storage of cells or packets awaiting service, and a first-in–first-out (FIFO) service discipline. Even in this simple case, the queueing behaviour depends on the type and mix of traffic being multiplexed through the buffer. So let's first look at the range of source models covered in the book, and then we'll summarize the queueing analysis results.

Source models

Model:	**negative exponential distribution**

Use:	inter-arrival times, service times, for cells, packets, bursts, flows, calls
Formula:	$\Pr\{\text{inter-arrival time} \leqslant t\} = F(t) = 1 - e^{-\lambda \cdot t}$
Parameters:	t – time
	λ – rate of arrivals, or rate of service
Location:	Chapter 6, page 83

Model:	**geometric distribution**

Use:	inter-arrival times, service times, for cells, packets, bursts, flows, calls
Formulas:	$\Pr\{k \text{ time slots between arrivals}\} = (1 - p)^{k-1} \cdot p$
	$\Pr\{\leqslant k \text{ time slots between arrivals}\} = 1 - (1 - p)^{k}$
Parameters:	k – time slots
	p – probability of an arrival, or end of service, in a time slot
Location:	Chapter 6, page 85

Model:	**Poisson distribution**

Use:	number of arrivals or amount of work, for octets, cells, packets, bursts, flows, calls
Formulas:	$\Pr\{k \text{ arrivals in time } T\} = \dfrac{(\lambda \cdot T)^{k}}{k!} \cdot e^{-\lambda \cdot T}$
Parameters:	T – time
	k – number of arrivals, or amount of work
	λ – rate of arrivals
Location:	Chapter 6, page 86

Model:	**binomial distribution**

Use:	number of arrivals (in time, or from a number of inputs) or amount of work, for octets, cells, packets, bursts, flows, calls
Formula:	$\Pr\{k \text{ arrivals in } N \text{ time slots}\} = \dfrac{N!}{(N - k)! \cdot k!} \cdot (1 - p)^{N-k} \cdot p^{k}$
Parameters:	k – number of arrivals, or amount of work
	p – probability of an arrival, in a time slot or from an input
	N – number of time slots, or number of inputs
Location:	Chapter 6, page 86

Model:	**Batch distribution**
Use:	number of arrivals, or amount of work, for octets, cells, packets, bursts, flows, calls
Formulas:	$a(0) = 1 - p$
	$a(1) = p \cdot b(1)$
	$a(2) = p \cdot b(2)$
	$\vdots$
	$a(k) = p \cdot b(k)$
	$\vdots$
	$a(M) = p \cdot b(M)$
Parameters:	k – number of arrivals
	p – probability there is a batch of arrivals in a time slot
	$b(k)$ – probability there are k arrivals in a batch (given that there is a batch in a time slot)
	M – maximum number of arrivals in batch
Location:	Chapter 6, page 88

Model:	**ON–OFF two-state**
Use:	rate of arrivals, for octets, cells, packets
Formulas:	$T_{on} = \dfrac{1}{R} \cdot \mathrm{E[on]}$
	$T_{off} = \dfrac{1}{C} \cdot \mathrm{E[off]}$
Parameters:	R – rate of arrivals
	$\mathrm{E[on]}$ – mean number of arrivals in ON state
	C – service rate, or rate of time-base
	$\mathrm{E[off]}$ – mean number of time units in OFF state
Location:	Chapter 6, page 91

Model:	**Pareto distribution**
Use:	number of arrivals, or amount of work, for octets, cells, packets, etc.
Formulas:	$\Pr\{X > x\} = \left(\dfrac{\delta}{x}\right)^{\alpha}$
	$F(x) = 1 - \left(\dfrac{\delta}{x}\right)^{\alpha}$
	$f(x) = \dfrac{\alpha}{\delta} \cdot \left(\dfrac{\delta}{x}\right)^{\alpha+1}$
	$\mathrm{E}[x] = \delta \cdot \dfrac{\alpha}{\alpha - 1}$

(continued overleaf)

Model:	**Pareto distribution**
Parameters:	δ – minimum amount of work
	x – number of arrivals, or amount of work
	α – power law decay
Location:	Chapter 17, page 289

Queueing behaviour

There are a number of basic queueing relationships which are true, regardless of the pattern of arrivals or of service, assuming that the buffer capacity is infinite (or that the loss is *very* low). For the basic FIFO queue, there is a wide range of queueing analyses that can be applied to both IP and ATM, according to the multiplexing scenario. These queueing relationships and analyses are summarized below.

Model:	**elementary relationships**
Use:	queues with infinite buffer capacity
Formulas:	$\rho = \lambda \cdot s$
	$w = \lambda \cdot t_w$ (known as Little's formula)
	$q = \lambda \cdot t_q$ (ditto)
	$t_q = t_w + s$
	$q = w + \rho$
Parameters:	λ – mean number of arrivals per unit time
	s – mean service time for each customer
	ρ – utilization; fraction of time the server is busy
	w – mean number of customers waiting to be served
	t_w – mean time a customer spends waiting for service
	q – mean number of customers in the system (waiting or being served)
	t_q – mean time a customer spends in the system
Location:	Chapter 4, page 61

Model:	**M/M/1**
Use:	classical continuous-time queueing model; NB: assumes variable-size customers, so more appropriate for IP, but has been used for ATM
Formulas:	$q = \dfrac{\rho}{1 - \rho}$

(*continued*)

Model:	**M/M/1**
	$$t_w = \frac{\rho \cdot s}{1 - \rho}$$
	$\Pr\{\text{system size} = x\} = (1 - \rho)\rho^x$
	$\Pr\{\text{system size} > x\} = \rho^{x+1}$
Parameters:	ρ – utilization; load (as fraction of service rate) offered to system
	q – mean number in the system (waiting or being served)
	t_w – mean time spent waiting for service
	x – buffer capacity in packets or cells
Location:	Chapter 4, page 62

Model:	**batch arrivals, deterministic service, infinite buffer capacity**
Use:	exact M/D/1, binomial/D/1, and arbitrary batch distributions – these can be applied to ATM, and to IP (with fixed packet sizes)
Formulas:	$E[a] = \rho$
	$s(0) = 1 - E[a]$
	$$s(k) = \frac{s(k-1) - s(0) \cdot a(k-1) - \sum_{i=1}^{k-1} s(i) \cdot a(k-i)}{a(0)}$$
	$\Pr\{U_d = 1\} = U_d(1) = s(0) + s(1)$
	$\Pr\{U_d = k\} = U_d(k) = s(k)$
	$$B_d(k) = \frac{1 - \sum_{i=0}^{k} a(i)}{E[a]}$$
	$$T_d(k) = \sum_{j=1}^{k} U_d(j) \cdot B_d(k-j)$$
	$$T_{d,n}(k) = \sum_{j=1}^{k} T_{d,n-1}(j) \cdot T_{d,1}(k-j)$$
	for M/D/1: $t_w = \dfrac{\rho \cdot s}{2 \cdot (1 - \rho)}$
Parameters:	$a(k)$ – probability there are k arrivals in a time slot
	ρ – utilization; load (as fraction of service rate) offered to system
	$E[a]$ – mean number of arrivals per time slot

(continued overleaf)

Model:	**batch arrivals, deterministic service, infinite buffer capacity**
	$s(k)$ – probability there are k in the system at the end of any slot
	$U_d(k)$ – probability there are k units of unfinished work in the buffer
	$B_d(k)$ – probability there are k arrivals ahead in arriving batch
	$T_d(k)$ – probability that an arrival experiences total delay of k
	$T_{d,n}(k)$ – probability that total delay through n buffers is k
	s – mean service time for each customer
	t_w – mean time spent waiting for service
Location:	Chapter 7, pages 100, 109, 110; and Chapter 4, page 66 (M/D/1 waiting time)

Model:	**batch arrivals, deterministic service, finite buffer capacity**
Use:	exact M/D/1, binomial/D/1, and arbitrary batch distributions – these can be applied to ATM, and to IP (with fixed packet sizes)
Formulas:	$A(k) = 1 - a(0) - a(1) - \cdots - a(k-1)$
	$u(0) = 1$
	$$u(k) = \frac{u(k-1) - a(k-1) - \sum_{i=1}^{k-1} u(i) \cdot a(k-i)}{a(0)}$$
	$u(X) = A(X)$
	$$s(0) = \frac{1}{\sum_{i=0}^{X} u(i)}$$
	$s(k) = s(0) \cdot u(k)$
	$$\text{CLP} = \frac{E[a] - (1 - s(0))}{E[a]}$$
Parameters:	$a(k)$ – probability there are k arrivals in a time slot
	$A(k)$ – probability there are at least k arrivals in a time slot
	$E[a]$ – mean number of arrivals per time slot
	$s(k)$ – probability there are k cells in the system at the end of any slot
	ρ – utilization; load (as fraction of service rate) offered to system
	CLP – probability of loss (whether cells or packets)
Location:	Chapter 7, page 105

Model:	**N·D/D/1**

Use: multiple constant-bit-rate (CBR) sources into deterministic server – this can be applied to ATM, and to IP (with fixed packet sizes)

Formulas:
$$Q(x) = \sum_{n=x+1}^{N} \left\{ \frac{N!}{n! \cdot (N-n)!} \cdot \left(\frac{n-x}{D} \right)^n \cdot \left[1 - \left(\frac{n-x}{D} \right) \right]^{N-n} \cdot \frac{D-N+x}{D-n+x} \right\}$$

Parameters: x – buffer capacity (in cells or packets)

                N – number of CBR sources

                D – period of CBR source (in service time slots)

                $Q(x)$ – probability that queue exceeds x (estimate for loss probability)

Location: Chapter 8, page 116

Model:	**M/D/1 heavy-traffic approximation**

Use: cell-scale queueing in ATM, basic packet queueing in IP (with fixed packet sizes); NB: below $\approx$80% load, underestimates loss

Formulas:
$$Q(x) = e^{-2 \cdot x \cdot \left(\frac{1-\rho}{\rho} \right)}$$

$$x = -\frac{1}{2} \cdot \ln(Q(x)) \cdot \left(\frac{\rho}{1-\rho} \right)$$

$$\rho = \frac{2 \cdot x}{2 \cdot x - \ln(Q(x))}$$

Parameters: x – buffer capacity (in cells or packets)

                ρ – utilization; load (as fraction of service rate) offered to system

                $Q(x)$ – probability that queue exceeds x (estimate for loss probability)

Location: Chapter 8, page 117

Model:	**N·D/D/1 heavy-traffic approximation**

Use: multiple constant-bit-rate (CBR) sources into deterministic server – this can be applied to ATM, and to IP (with fixed packet sizes); NB: below $\approx$80% load, underestimates performance

Formulas:
$$Q(x) = e^{-2 \cdot x \cdot \left(\frac{x}{N} + \frac{1-\rho}{\rho} \right)}$$

Parameters: x – buffer capacity (in cells or packets)

(continued overleaf)

Model:	**N·D/D/1 heavy-traffic approximation**
	ρ – utilization; load (as fraction of service rate) offered to system
	N – number of CBR sources
	$Q(x)$ – probability that queue exceeds x (estimate for loss probability)
Location:	Chapter 8, page 120

Model:	**Geo/Geo/1**
Use:	basic discrete-time queueing model for IP (variable-size packets)
Formulas:	$s(0) = 1 - \dfrac{p}{q}$
	$s(k) = \left(1 - \dfrac{p}{q}\right) \cdot \dfrac{p}{1-q} \cdot \left(\dfrac{1-q}{1-p}\right)^{k}$
	$Q(k) = \dfrac{p}{q} \cdot \left(\dfrac{1-q}{1-p}\right)^{k}$
	$Q(x) = \dfrac{p}{q} \cdot \left(\dfrac{1-q}{1-p}\right)^{x/q}$
Parameters:	q – probability a packet completes service at the end of an octet slot
	p – probability a packet arrives in an octet slot
	$s(k)$ – probability there are k octets in system
	$Q(k)$ – probability that queue exceeds k octets
	$Q(x)$ – probability that queue exceeds x packets
Location:	Chapter 14, page 232

Model:	**excess-rate, Geometrically Approximated Poisson Process (GAPP), M/D/1**
Use:	accurate approximation to M/D/1 – can be applied to ATM, and to IP (with fixed packet sizes)
Formulas:	$p(k) = \left(1 - \dfrac{\lambda \cdot e^{\lambda} - e^{\lambda} - \lambda^{2} + \lambda + e^{-\lambda}}{\lambda - 1 + e^{-\lambda}}\right) \cdot \left[\dfrac{\lambda \cdot e^{\lambda} - e^{\lambda} - \lambda^{2} + \lambda + e^{-\lambda}}{\lambda - 1 + e^{-\lambda}}\right]^{k}$
	$Q(k) = \left[\dfrac{\lambda \cdot e^{\lambda} - e^{\lambda} - \lambda^{2} + \lambda + e^{-\lambda}}{\lambda - 1 + e^{-\lambda}}\right]^{k+1}$
Parameters:	λ – arrival rate of Poisson process
	$p(k)$ – probability an arriving excess-rate cell/packet finds k in the system

(continued)

Model:	**excess-rate, Geometrically Approximated Poisson Process (GAPP), M/D/1**
	$Q(k)$ – probability an arriving excess-rate cell/packet finds more than k in the system
Location:	Chapter 14, page 245

Model:	**excess-rate GAPP analysis for bi-modal service distributions**
Use:	accurate approximation to M/bi-modal/1 – suitable for IP, with bi-modal distribution to model short and long packets
Formulas:	$E[a] = \lambda \cdot (p_s + (1 - p_s) \cdot n)$

$$a(0) = p_s \cdot e^{-\lambda} + (1 - p_s) \cdot e^{-n \cdot \lambda}$$

$$a(1) = p_s \cdot \lambda \cdot e^{-\lambda} + (1 - p_s) \cdot n \cdot \lambda \cdot e^{-n \cdot \lambda}$$

$$p(k) = \left(1 - \frac{E[a] \cdot (1 - a(1)) - 1 + a(1) + (a(0))^2}{a(0) \cdot (E[a] - 1 + a(0))} \right)$$

$$\cdot \left[\frac{E[a] \cdot (1 - a(1)) - 1 + a(1) + (a(0))^2}{a(0) \cdot (E[a] - 1 + a(0))} \right]^k$$

$$Q(k) = \left[\frac{E[a] \cdot (1 - a(1)) - 1 + a(1) + (a(0))^2}{a(0) \cdot (E[a] - 1 + a(0))} \right]^{k+1}$$

Parameters:	$a(k)$ – probability there are k arrivals in a packet service time
	$E[a]$ – mean number of arrivals per packet service time
	λ – packet arrival rate of Poisson process (i.e. per time unit = short packet)
	p_s – proportion of short packets
	n – length of long packets (multiple of short packet)
	$p(k)$ – probability an arriving excess-rate packet finds k in the system
	$Q(k)$ – probability an arriving excess-rate packet finds more than k in the system
Location:	Chapter 14, page 249

Model:	**excess-rate GAPP analysis for M/G/1**
Use:	accurate approximation to M/G/1 – suitable for IP, with general service time distribution to model variable-length packets

(continued overleaf)

Model:	**excess-rate GAPP analysis for M/G/1**

Formulas:

$$E[a] = \lambda \cdot \sum_{i=1}^{\infty} g(i) = \lambda$$

$$a(0) = \sum_{i=1}^{\infty} g(i) \cdot e^{-i \cdot \lambda}$$

$$a(1) = \sum_{i=1}^{\infty} g(i) \cdot i \cdot \lambda \cdot e^{-i \cdot \lambda}$$

$$p(k) = \left(1 - \frac{E[a] \cdot (1 - a(1)) - 1 + a(1) + (a(0))^2}{a(0) \cdot (E[a] - 1 + a(0))} \right)$$

$$\cdot \left[\frac{E[a] \cdot (1 - a(1)) - 1 + a(1) + (a(0))^2}{a(0) \cdot (E[a] - 1 + a(0))} \right]^k$$

$$Q(k) = \left[\frac{E[a] \cdot (1 - a(1)) - 1 + a(1) + (a(0))^2}{a(0) \cdot (E[a] - 1 + a(0))} \right]^{k+1}$$

Parameters: $A(k)$ – probability there are k arrivals in a packet service time

$E[a]$ – mean number of arrivals per packet service time

λ – packet arrival rate of Poisson process (i.e. per unit time)

$g(k)$ – probability a packet requires k units of time to be served

$p(k)$ – probability an arriving excess-rate packet finds k in the system

$Q(k)$ – probability an arriving excess-rate packet finds more than k in the system

Location: Chapter 14, page 249

Model:	**ON–OFF/D/1/K**

Use: basic continuous-time queueing model for IP or ATM, suitable for per-flow or per-VC scenarios

Formulas:

$$\alpha = \frac{T_{on}}{T_{on} + T_{off}}$$

$$CLP_{excess\text{-}rate} = \frac{(C - \alpha \cdot R) \cdot e^{\left(\frac{-X \cdot (C - \alpha \cdot R)}{T_{on} \cdot (1 - \alpha) \cdot (R - C) \cdot C} \right)}}{(1 - \alpha) \cdot C - \alpha \cdot (R - C) \cdot e^{\left(\frac{-X \cdot (C - \alpha \cdot R)}{T_{on} \cdot (1 - \alpha) \cdot (R - C) \cdot C} \right)}}$$

$$CLP = \frac{R - C}{R} \cdot CLP_{excess\text{-}rate}$$

Parameters: R – ON rate

C – service rate of buffer

(continued)

Model:	**ON–OFF/D/1/K**

X – buffer capacity in cells/packets

T_{on} – mean duration in ON state

T_{off} – mean duration in OFF state

α – activity factor of source (probability of being ON)

CLP – loss probability

Location: Chapter 9, page 130

Model:	**ON–OFF/D/1/K**

Use: basic discrete-time queueing model for IP or ATM, suitable for per-flow or per-VC scenarios

Formulas:

$$a = 1 - \frac{1}{T_{on} \cdot (R - C)}$$

$$s = 1 - \frac{1}{T_{off} \cdot C}$$

$$p(X) = \frac{1}{1 + \left(\left(\frac{s}{a} \right)^X - 1 \right) \cdot \left(\frac{1 - a}{s - a} \right)}$$

$$\mathrm{CLP} = \frac{R - C}{R} \cdot \mathrm{CLP}_{\text{excess-rate}} = \frac{R - C}{R} \cdot p(X)$$

Parameters: R – ON rate

C – service rate of buffer

X – buffer capacity in cells/packets

T_{on} – mean duration in ON state

T_{off} – mean duration in OFF state

$p(k)$ = probability an excess-rate arrival finds k in the buffer

CLP – loss probability

Location: Chapter 9, page 136

Model:	**multiple ON–OFF sources – bufferless analysis**

Use: burst-scale loss model for IP or ATM – for delay-sensitive traffic, or, combined with burst-scale delay analysis, for delay-insensitive traffic

Formulas:

$$\alpha = \frac{m}{h} = \frac{T_{on}}{T_{on} + T_{off}}$$

$$N_0 = \frac{C}{h}$$

(*continued overleaf*)

Model:	**multiple ON–OFF sources – bufferless analysis**

$$p_n = \frac{N!}{n! \cdot (N-n)!} \cdot \alpha^n \cdot (1-\alpha)^{N-n}$$

$$\Pr\{\text{cell needs buffer}\} = \frac{\displaystyle\sum_{n=\lceil N_0 \rceil}^{N} p_n \cdot (n - N_0)}{N \cdot \alpha}$$

Parameters:	m – mean rate of single source
	h – ON rate of single source
	T_{on} – mean duration in ON state for single source
	T_{off} – mean duration in OFF state for single source
	α – activity factor of single source (probability of being ON)
	C – service rate of buffer
	N_0 – minimum number of active sources for burst-scale queueing
	N – total number of ON–OFF sources being multiplexed
	p_n = probability that n sources are active
	$\Pr\{\text{cell needs buffer}\}$ – estimate of loss probability
Location:	Chapter 9, page 141

Model:	**multiple ON–OFF sources – *approximate* bufferless analysis**
Use:	burst-scale loss model for IP or ATM – for delay-sensitive traffic, or, combined with burst-scale delay analysis, for delay-insensitive traffic

Formulas:

$$\rho = \frac{N \cdot m}{C}$$

$$N_0 = \frac{C}{h}$$

$$\Pr\{\text{cell needs buffer}\} \approx \frac{1}{(1-\rho)^2 \cdot N_0} \cdot \frac{(\rho \cdot N_0)^{\lfloor N_0 \rfloor}}{\lfloor N_0 \rfloor!} \cdot e^{-\rho \cdot N_0}$$

Parameters:	m – mean rate of single source
	h – ON rate of single source
	C – service rate of buffer
	N – total number of ON–OFF sources being multiplexed
	ρ – offered load as fraction of service rate
	N_0 – minimum number of active sources for burst-scale queueing
	$\Pr\{\text{cell needs buffer}\}$ – estimate of loss probability
Location:	Chapter 9, page 142

Model:	**multiple ON–OFF sources – burst-scale delay analysis**

Use:	burst-scale queueing model for IP or ATM – combined with burst-scale loss (bufferless) analysis, for delay-insensitive traffic
Formulas:	$\lambda = \dfrac{N}{T_{on} + T_{off}}$
	$b = T_{on} \cdot h$
	$\rho = \dfrac{b \cdot \lambda}{C}$
	$N_0 = \dfrac{C}{h}$
	$\text{CLP}_{\text{excess-rate}} = e^{-\left[N_0 \cdot \frac{X}{b} \cdot \frac{(1-\rho)^3}{4 \cdot \rho + 1} \right]}$
Parameters:	N – total number of ON–OFF sources being multiplexed
	T_{on} – mean duration in ON state for single source
	T_{off} – mean duration in OFF state for single source
	h – ON rate of single source
	C – service rate of buffer
	λ – number of bursts arriving per unit time
	b – mean number of cells/packets per burst
	ρ – offered load as fraction of service rate
	N_0 – minimum number of active sources for burst-scale queueing
	$\text{CLP}_{\text{excess-rate}}$ – excess-rate loss probability, i.e. conditioned on the probability that the cell/packet needs a buffer
Location:	Chapter 9, page 146

Model:	**multiple ON–OFF sources – excess-rate analysis**

Use:	combined burst-scale loss and delay analysis – suitable for IP and ATM scenarios with multiple flows (e.g. RSVP), or variable-bit-rate (VBR) traffic (e.g. SBR/VBR transfer capability)
Formulas:	$N_0 = \dfrac{C}{h}$
	$A = \dfrac{A_p}{h}$
	$D = \left\{ \dfrac{A^{N_0}}{N_0!} \cdot \left(\dfrac{N_0}{N_0 - A} \right) \right\} \Big/ \left\{ \displaystyle\sum_{r=0}^{N_0-1} \dfrac{A^r}{r!} + \dfrac{A^{N_0}}{N_0!} \cdot \left(\dfrac{N_0}{N_0 - A} \right) \right\}$
	$T(on) = \dfrac{h \cdot T_{on}}{C - A_p}$

(continued overleaf)

Model:	**multiple ON–OFF sources – excess-rate analysis**

$$R_{on} = C + h \cdot \frac{A_p}{C - A_p}$$

$$T(\textit{off}) = T(\textit{on}) \cdot \frac{1 - D}{D}$$

$$R_{\textit{off}} = \frac{A_p - D \cdot R_{on}}{1 - D}$$

$$Q(x) = \frac{h \cdot D}{C - A_p} \cdot \left(\frac{1 - \dfrac{1}{T(\textit{on}) \cdot (R_{on} - C)}}{1 - \dfrac{1}{T(\textit{off}) \cdot (C - R_{\textit{off}})}} \right)^{x+1}$$

Parameters:

h – ON rate of flow in packet/s or cell/s

C – service rate of buffer

N_0 – minimum number of active sources for burst-scale queueing

A_p – overall mean load in packet/s or cell/s

A – offered traffic in packet flows (equivalent to erlang occupancy of circuits, each circuit of rate h)

D – probability of a packet flow waiting, i.e. of being in excess rate state

T_{on} – mean duration of flow

$T(\textit{on})$ – mean duration in excess-rate ON state

R_{on} – mean input rate to buffer when in excess-rate ON state

$T(\textit{off})$ – mean duration in underload OFF state

$R_{\textit{off}}$ – mean input rate to buffer when in underload OFF state

$Q(x)$ – queue overflow probability for buffer size of x packets (estimate for loss probability)

Location: Chapter 15, page 261

Model:	**Geo/Pareto/1**

Use: discrete-time queueing model for LRD (long-range dependence) traffic in IP or ATM – can be viewed as batch arrival process with Pareto-distributed number of packets, or geometric arrivals with Pareto-distributed service times

Formulas:

$$b(1) = F(1.5) - F(1) = 1 - \left(\frac{1}{1.5} \right)^\alpha$$

$$b(x) = F(x + 0.5) - F(x - 0.5) = \left(\frac{1}{x - 0.5} \right)^\alpha - \left(\frac{1}{x + 0.5} \right)^\alpha$$

$$B = \frac{\alpha}{\alpha - 1}$$

(continued)

Model:	**Geo/Pareto/1**

$$q = \frac{\rho}{B}$$

$$a(0) = 1 - q$$

$$a(1) = q \cdot b(1)$$

$$a(2) = q \cdot b(2)$$

$$\vdots$$

$$a(k) = q \cdot b(k)$$

$$E[a] = \rho$$

$$s(0) = 1 - E[a]$$

$$s(k) = \frac{s(k-1) - s(0) \cdot a(k-1) - \sum_{i=1}^{k-1} s(i) \cdot a(k-i)}{a(0)}$$

Parameters: x – number of arrivals, or amount of work

a – power-law decay

$b(x)$ – probability that Pareto batch is of size x packets

B – mean batch size in packets

ρ – mean number of packets arriving per time unit

q – probability that a batch arrives in a time unit

$a(k)$ – probability there are k arrivals in a time unit

$E[a]$ – mean number of arrivals per time unit

$s(k)$ – probability there are k in the system at the end of any time unit

Location: Chapter 17, page 293

Model:	**Geo/truncated Pareto/1**

Use: discrete-time queueing model for LRD traffic in IP or ATM – NB: truncated Pareto distribution limits range of time scales of bursty behaviour, giving more realistic LRD traffic model

Formulas: $$b(x) = \begin{cases} \left(1 - \left(\frac{1}{0.5}\right)^{\alpha}\right) \Big/ \left(1 - \left(\frac{1}{X+0.5}\right)^{\alpha}\right) & x = 1 \\[2ex] \left(\left(\frac{1}{x-0.5}\right)^{\alpha} - \left(\frac{1}{x+0.5}\right)^{\alpha}\right) \Big/ \left(1 - \left(\frac{1}{X+0.5}\right)^{\alpha}\right) & X \geqslant x > 1 \\[2ex] 0 & x > X \end{cases}$$

$$B = \sum_{x} x \cdot b(x)$$

(continued overleaf)

Model: **Geo/truncated Pareto/1**

$$q = \frac{\rho}{B}$$

$$a(0) = 1 - q$$

$$a(1) = q \cdot b(1)$$

$$a(2) = q \cdot b(2)$$

$$\vdots$$

$$a(k) = q \cdot b(k)$$

$$E[a] = \rho$$

$$s(0) = 1 - E[a]$$

$$s(k) = \frac{s(k-1) - s(0) \cdot a(k-1) - \sum_{i=1}^{k-1} s(i) \cdot a(k-i)}{a(0)}$$

Parameters: x – number of arrivals, or amount of work

α – power-law decay

$b(x)$ – probability that Pareto batch is of size x packets

B – mean batch size in packets

ρ – mean number of packets arriving per time unit

q – probability that a batch arrives in a time unit

$a(k)$ – probability there are k arrivals in a time unit

$E[a]$ – mean number of arrivals per time unit

$s(k)$ – probability there are k in the system at the end of any time unit

Location: Chapter 17, page 298

COPING WITH MULTI-SERVICE REQUIREMENTS: DIFFERENTIATED PERFORMANCE

A FIFO discipline does not allow different performance requirements to be guaranteed by the network – in best-effort IP all traffic suffers similar delay and loss, and in ATM the most stringent requirement limits the admissible load. The solution is to manage the buffer, both on entry and at the exit – this involves policies for partitioning and sharing the buffer space and server capacity (e.g. per-flow/per-VC queueing), packet and cell discard mechanisms, and queue scheduling (such as precedence queueing and weighted fair queueing).

Buffer sharing and partitioning

With per-flow/per-VC queueing and weighted fair queueing, each virtual buffer can be modelled as having its own server capacity and buffer space – thus any of the analysis methods for FIFO queues (see previous section) can be applied, as appropriate to the multiplexing scenario, and traffic source(s). Typically these will give a decay rate for each virtual buffer, which can then be used, along with the performance requirement, to assess the partitioning of buffer space.

There is clearly benefit in partitioning to maintain different performance guarantees for a variety of service types sharing an output port. However, the cost of partitioning is that it is not optimal when considering the overall loss situation at an output port: the loss of a cell or packet from a full virtual buffer may not be necessary if buffer space is shared. Indeed, buffer space can be shared across multiple output ports. The results for both partitioning and sharing are summarized below.

Model:	**buffer-space partitioning analysis**
Use:	to allocate space to virtual buffers according to performance requirements, and decay rate (obtained from queueing analysis)
Formulas:	$S_1 \cdot dr_1^{X_1} = S_2 \cdot dr_2^{X_2} = \cdots = S_j \cdot dr_j^{X_j} = \cdots = S_V \cdot dr_V^{X_V}$
	$$X = \sum_{j=1}^{V} X_j$$
	$$X_i = \frac{X + \sum_{j=1}^{V} \left(\dfrac{\log(S_j)}{\log(dr_j)} \right)}{\log(dr_i) \cdot \sum_{j=1}^{V} \left(\dfrac{1}{\log(dr_j)} \right)} - \frac{\log(S_i)}{\log(dr_i)}$$
Parameters:	X – total buffer space available
	X_i – buffer space allocation for virtual buffer i
	S_i – overflow probability scaling factor for virtual buffer i
	dr_i – decay rate obtained from queueing analysis of virtual buffer i
Location:	Chapter 16, page 278

Model:	**shared buffer analysis**
Use:	to assess the performance improvement when sharing buffer space across multiple output buffers, using decay rate (obtained from queueing analysis)
Formulas:	$p(k) = (1 - d_r) \cdot (d_r)^k$

(continued overleaf)

Model:	**shared buffer analysis**

$$P_N(k) = \sum_{j=0}^{k} P_{N-1}(j) \cdot P_1(k-j)$$

$$Q_N(k) = 1 - \sum_{j=0}^{k} P_N(j)$$

negative binomial approximation:

$$Q_N(k-1) \approx {}^{k+N-1}C_{N-1} \cdot (d_r)^k \cdot (1-d_r)^{N-1} \text{ or}$$

$$Q_N(k-1) \approx e^{\{(k+N-1)\cdot\ln(k+N-1)-k\cdot\ln(k)-(N-1)\cdot\ln(N-1)+k\cdot\ln(d_r)+(N-1)\cdot\ln(1-d_r)\}}$$

Parameters: d_r – decay rate in individual buffer

$p(k)$ – queue state probability for single (virtual) buffer, i.e. probability that individual buffer has k cells/packets

$P_N(k)$ – autoconvolution for N buffers sharing space, i.e. probability that shared space has k cells/packets

$Q_N(k)$ – overflow probability from shared buffer

Location: Chapter 16, page 280

Cell and packet discard mechanisms

In ATM, the space priority mechanism enables a FIFO buffer to provide two different levels of loss performance, based on a threshold level for the queue size. Combined with virtual buffers, it is possible to provide this type of loss differentiation to traffic classes, or indeed within a VC, to the different CLP 0/CLP 1 flows.

In IP, random early detection (RED) provides a probabilistic discard mechanism with the aim of alleviating congestion, and avoiding it if possible – addressing the global synchronization problems associated with multiple TCP connections. RED also uses a threshold mechanism, but unlike in ATM, this works in conjunction with the average queue size.

Analysis for both of these discard mechanisms is summarized below.

Model:	**M/D/1/K with partial buffer sharing (PBS)**
Use:	basic discrete-time queueing model for analysing PBS in ATM
Formulas:	$a = a_h + a_l$

$$a(k) = \frac{a^k}{k!} \cdot e^{-a}$$

$$a_h(k) = \frac{a_h{}^k}{k!} \cdot e^{-a_h}$$

(continued)

Model:	**M/D/1/K with partial buffer sharing (PBS)**

$$A(k) = 1 - \sum_{j=0}^{k-1} a(j)$$

$$A_h(k) = 1 - \sum_{j=0}^{k-1} a_h(j)$$

$$a'(m, n) = \sum_{i=m+n}^{\infty} \left[a(i) \cdot \frac{(i-m)!}{n! \cdot (i-m-n)!} \cdot \left(\frac{a_h}{a}\right)^n \cdot \left(\frac{a_l}{a}\right)^{i-m-n} \right]$$

$$A'(m, n) = 1 - \sum_{i=0}^{m+n-1} a(i) - \sum_{i=0}^{\infty} \left[a(m+n+i) \right.$$
$$\left. \cdot \sum_{j=0}^{n-1} \left\{ \frac{(n+i)!}{j! \cdot (n+i-j)!} \cdot \left(\frac{a_h}{a}\right)^j \cdot \left(\frac{a_l}{a}\right)^{n+i-j} \right\} \right]$$

$$u(0) = 1$$

$$u(1) = \frac{(1 - a(0))}{a(0)}$$

below the threshold:

$$u(k) = \frac{A(k) + \sum_{i=1}^{k-1} u(i) \cdot A(k-i+1)}{a(0)}$$

$$u(M) = \frac{A(M) + \sum_{i=1}^{M-1} u(i) \cdot A'(M-i, 1)}{a_h(0)}$$

above the threshold:

$$u(k) = \frac{A'(M, k-M) + \sum_{i=1}^{M-1} \{u(i) \cdot A'(M-i, k-M+1)\} + \sum_{i=M}^{k-1} \{u(i) \cdot A_h(k-i+1)\}}{a_h(0)}$$

$$u(X) = \frac{A'(M, X-M)}{a_h(0)}$$

$$s(0) = \frac{1}{\sum_{i=0}^{X} u(i)}$$

$$s(k) = s(0) \cdot u(k)$$

(continued overleaf)

Model:	**M/D/1/K with partial buffer sharing (PBS)**

$$\text{CLP} = \frac{a_l + a_h - (1 - s(0))}{a_l + a_h}$$

$$l_h(j) = \sum_{i=0}^{M-1} s(i) \cdot a'(M - i, X - M + j) + \sum_{i=M}^{X} s(i) \cdot a_h(X - i + j)$$

$$\text{CLP}_h = \frac{\sum_j j \cdot l_h(j)}{a_h}$$

$$l_l(j) = \sum_{i=0}^{M-1} \left[s(i) \cdot \sum_{r=M-i+j}^{\infty} a(r) \cdot \frac{(r - (M - i))!}{(r - (M - i) - j)! \cdot j!} \cdot \left(\frac{a_h}{a} \right)^{r-(M-i)-j} \right.$$
$$\left. \cdot \left(\frac{a_l}{a} \right)^j \right] + \sum_{i=M}^{X} s(i) \cdot a_l(j)$$

$$\text{CLP}_l = \frac{\sum_j j \cdot l_l(j)}{a_l}$$

Parameters:
a – mean arrival rate of both high- and low-priority arrivals

a_l – mean arrival rate of low-priority arrivals

a_h – mean arrival rate of high-priority arrivals

$a(k)$ – probability there are k arrivals in one time slot

$a_l(k)$ – probability there are k low-priority arrivals in one time slot

$a_h(k)$ – probability there are k high-priority arrivals in one time slot

$A(k)$ – probability at least k cells arrive in one time slot

$A_h(k)$ – probability at least k high-priority cells arrive in one time slot

$a'(m, n)$ – probability that m cells of either low or high priority are admitted, up to the threshold, and a further n high-priority cells are admitted above the threshold

$A'(m, n)$ – probability that m cells of either low or high priority are admitted, up to the threshold, and at least a further n high-priority cells are admitted above the threshold

$s(k)$ – probability there are k cells in the system

$l_h(j)$ – probability that j high-priority cells are lost in a time slot

$l_l(j)$ – probability that j low-priority cells are lost in a time slot

M – PBS threshold (cells)

X – buffer capacity in cells

CLP – overall cell loss probability

CLP_h – cell loss probability for high-priority cells

CLP_l – cell loss probability for low-priority cells

Location: Chapter 13, page 207

Model:	**M/D/1/K with PBS – margin between high- and low-priority CLP**
Use:	rule of thumb for configuring PBS threshold, given high-priority load
Formulas:	$\text{CLP}_{\text{margin}} = 10^{-(X-M)}$ for $a_h = 0.25$
	$\text{CLP}_{\text{margin}} = 10^{-2 \cdot (X-M)}$ for $a_h = 0.04$
Parameters:	a_h – mean arrival rate of high-priority arrivals
	M – PBS threshold (cells)
	X – buffer capacity in cells
	$\text{CLP}_{\text{margin}}$ – difference between high- and low-priority cell loss probability
Location:	Chapter 16, page 218

Model:	**threshold indicators for random early detection (RED) mechanism**
Use:	mean queue size based on decay rate analysis, to aid in configuring EWMA thresholds: θ_{min} and θ_{max}
Formulas:	$p(k) = (1 - d_r) \cdot (d_r)^k$
	$q = \dfrac{d_r}{1 - d_r}$
Parameters:	d_r – decay rate based on appropriate queueing analysis (see earlier, e.g. use both basic packet-scale and burst-scale decay rates)
	$p(k)$ – probability that queue contains k packets
	q – mean queue size
Location:	Chapter 16, page 270

Queue scheduling mechanisms

Whether in IP or in ATM, queue scheduling mechanisms involve partitioning the service capacity of the output buffer. Weighted fair queueing allocates a specified proportion of the server to each virtual buffer – to analyse this, each virtual buffer can be treated as an independent FIFO queue with a fixed service capacity equal to its allocation. Thus any of the analysis methods for FIFO queues (see previous section) can be applied. Precedence queueing, i.e. time priority, requires specific analysis, because the partitions between different priority levels are not fixed. The amount of service capacity seen by lower-priority traffic depends on how much has been used by higher-priority traffic. The queueing analysis for time priorities is summarized below.

Model:	**M/D/1 with time priorities – mean value analysis**
Use:	basic discrete-time queueing model for analysing time priorities in ATM – also applies to IP with fixed packet sizes

Formulas:

$$w_1 = \frac{a_1 + a_2}{2 \cdot (1 - a_1)}$$

$$w_2 = \frac{w_1 \cdot a_1 + \dfrac{a_1 + a_2}{2}}{1 - a_1 - a_2}$$

Parameters: a_i – mean arrival rate of priority-i arrivals (highest priority is 1)
                   w_i – mean waiting time of priority-i arrivals

Location: Chapter 13, page 219

Model:	**M/D/1 with time priorities – waiting-time distribution analysis**
Use:	basic discrete-time queueing model for analysing time priorities in ATM – also applies to IP with fixed packet sizes

Formulas:

$$E[a] = \rho$$

$$s(0) = 1 - E[a]$$

$$s(k) = \frac{s(k-1) - s(0) \cdot a(k-1) - \displaystyle\sum_{i=1}^{k-1} s(i) \cdot a(k-i)}{a(0)}$$

$$a_1(k) = \frac{a_1^{\,k}}{k!} \cdot e^{-a_1}$$

$$a_2(k) = \frac{a_2^{\,k}}{k!} \cdot e^{-a_2}$$

$$b(k) = \frac{1 - \displaystyle\sum_{i=0}^{k} a_2(i)}{a_2}$$

$$u(0) = s(0) + s(1)$$

$$u(k) = s(k+1) \quad \text{for } k > 0$$

$$v(k) = \sum_{i=0}^{k} \left[u(k-i) \cdot \sum_{j=0}^{i} b(j) \cdot a_1(i-j) \right]$$

$$a_1(k, x) = \frac{a_1^{\,x \cdot k}}{x!} \cdot e^{-k \cdot a_1}$$

$$w(0) = v(0)$$

$$w(k) = \frac{\displaystyle\sum_{i=1}^{k} v(i) \cdot a_1(k-i, k) \cdot i}{k} \qquad \text{for } k > 0$$

(continued)

Model:	**M/D/1 with time priorities – waiting-time distribution analysis**
Parameters:	$E[a]$ – mean arrival rate of both high- and low-priority arrivals
	a_i – mean arrival rate of priority-i arrivals (highest priority is 1)
	$a(k)$ – probability there are k arrivals in one time slot
	$a_i(k)$ – probability there are k arrivals of priority i in one time slot
	$s(k)$ – probability of k cells/packets in queue
	$u(k)$ – probability of k units of unfinished work in queue
	$b(k)$ – probability there are k arrivals ahead in arriving batch
	$v(k)$ – virtual waiting time distribution
	$a_1(k, x)$ – probability that k high-priority cells arrive in x time slots
	$w(k) = \Pr\{$a priority-2 cell must wait k time slots before it enters service$\}$
Location:	Chapter 13, page 222

FLOWS, CONNECTIONS AND AGGREGATES

To provide end-to-end performance guarantees, traffic contracts must be established, resources reserved, and traffic flows monitored to ensure compliance with the contract. At the heart of these functions is the assessment of performance resulting from traffic flowing over network resources, i.e. various forms of queueing analysis.

Admission control mechanisms

Whether in ATM or in (QoS-aware) IP, an admission control function has to assess whether or not a new connection or flow can be admitted. This function must take into account the commitments that the network is currently supporting and the resources available for any new flows. To do this, it needs to be aware of the structure and configuration of the buffers (i.e. the scheduling, discard and partitioning mechanisms), and make an appropriate assessment of whether the requested performance can be guaranteed. This requires queueing analysis, as summarized in previous sections in this chapter, in a form suitable for fast, accurate responses. Examples of these forms are summarized below.

Model:	**admissible load based on N·D/D/1 analysis**
Use:	admission control in ATM (for DBR, i.e. cell-scale component) or IP (e.g. voice-over-IP with fixed packet sizes) for constant-bit-rate traffic

(continued overleaf)

Model:	**admissible load based on N·D/D/1 analysis**

Formulas: If

$$n + 1 \leqslant -\frac{2 \cdot x^2}{\ln\left(\min_{i=1 \to n+1}(\text{CLP}_i)\right)}$$

then accept if

$$\frac{h_{n+1}}{C} + \sum_{i=1}^{n} \frac{h_i}{C} \leqslant 1$$

If

$$n + 1 > -\frac{2 \cdot x^2}{\ln\left(\min_{i=1 \to n+1}(\text{CLP}_i)\right)}$$

then accept if $\dfrac{h_{n+1}}{C} + \sum_{i=1}^{n} \dfrac{h_i}{C}$

$$\leqslant \frac{2 \cdot x \cdot (n+1)}{2 \cdot x \cdot (n+1) - \left[2 \cdot x^2 + (n+1) \cdot \ln\left(\min_{i=1 \to n+1}(\text{CLP}_i)\right)\right]}$$

The above is based on heavy-traffic approximation; for alternative based on more accurate N·D/D/1 analysis, see Tables 10.2 and 10.3

Parameters: h_i – fixed arrival rate of ith flow or connection

CLP$_i$ – loss requirement of ith flow or connection

n – number of existing flows or connections

C – service rate of (virtual) buffer

x – buffer capacity in packets or cells

Location: Chapter 10, page 153

Model:	**admissible load based on M/D/1 analysis**

Use: admission control in ATM (for DBR, and cell-scale component in SBR) or IP (e.g. voice-over-IP with fixed packet sizes)

Formulas: $\dfrac{h_{n+1}}{C} + \sum_{i=1}^{n} \dfrac{h_i}{C} \leqslant \dfrac{2 \cdot x}{2 \cdot x - \ln\left(\min_{i=1 \to n+1}(\text{CLP}_i)\right)}$

right-hand side of inequality test is based on heavy traffic approximation; for alternative based on exact M/D/1 analysis, see Table 10.1

Parameters: h_i – fixed arrival rate of ith flow or connection

CLP$_i$ – loss requirement of ith flow or connection

n – number of existing flows or connections

(continued)

Model:	**admissible load based on M/D/1 analysis**

C – service rate of (virtual) buffer

x – buffer capacity in packets or cells

Location: Chapter 10, page 152

Model:	**admissible load based on burst-scale loss analysis**

Use: admission control in ATM (burst-scale component in SBR) or IP (e.g. voice-over-IP with silence suppression) for delay-sensitive VBR traffic

Formulas:

$$N_0 = \frac{C}{h}$$

$$\frac{m_{n+1}}{C} + \sum_{i=1}^{n} \frac{m_i}{C} \leqslant \rho(\mathrm{CLP}, N_0)$$

where $\rho\,(\mathrm{CLP}, N_0)$ is based on e.g.

$$\mathrm{CLP} \approx \frac{1}{(1-\rho)^2 \cdot N_0} \cdot \frac{(\rho \cdot N_0)^{\lfloor N_0 \rfloor}}{\lfloor N_0 \rfloor!} \cdot e^{-\rho \cdot N_0}$$

and can be found in Table 10.4

Parameters: m_i – mean arrival rate of ith flow or connection

       CLP – most stringent loss requirement of all flows or connections

       n – number of existing flows or connections

       C – service rate of (virtual) buffer

       h – maximum individual peak arrival rate of all flows or connections

       N_0 – minimum number of active sources at maximum peak rate for burst-scale queueing

       ρ – admissible load

Location: Chapter 10, page 159

Model:	**admissible load based on burst-scale delay analysis**

Use: admission control in ATM (burst-scale delay component in SBR) or IP, for delay-insensitive VBR traffic

Formulas: Table 10.5 is based on burst-scale delay analysis:

$$\mathrm{CLP}_{\text{excess-rate}} = e^{-\left[N_0 \cdot \frac{X}{b} \cdot \frac{(1-\rho)^3}{4 \cdot \rho + 1}\right]}$$

and should be used in conjunction with burst-scale loss, Table 10.4

(continued overleaf)

Model:	**admissible load based on burst-scale delay analysis**
Parameters:	b – mean burst duration rate (maximum over all flows or connections)
	X – buffer capacity in cells or packets
	N_0 – minimum number of active sources at maximum peak rate for burst-scale queueing
	$CLP_{excess-rate}$ – loss component attributed to burst-scale delay analysis
	ρ – admissible load
Location:	Chapter 10, page 162

Policing mechanisms

Once admitted, the flow of packets or cells is monitored (to ensure compliance with the traffic contract) by a policing function, usually a token, or leaky, bucket. In IP, the token bucket is typically integrated into the queue scheduling mechanism, whereas in ATM, the leaky bucket is normally a separate function on entry to the network. Both can be assessed using a variety of forms of queueing analysis (see earlier), in order to know how to configure the buckets appropriately. Some examples of analysis are summarized below.

Model:	**worst-case analysis of leaky bucket for CDV tolerances**
Use:	allowance for impact on load of CDV tolerance in contract for CBR traffic using DBR transfer capability in ATM
Formulas:	$$MBS = 1 + \left\lfloor \frac{\tau}{T - \Delta} \right\rfloor = 1 + \left\lfloor \frac{\tau/\Delta}{D - 1} \right\rfloor$$ $$\rho = \frac{2 \cdot \left(\dfrac{X}{MBS} + D \right)}{D \cdot \left(2 - \dfrac{MBS \cdot \ln(CLP)}{X} \right)}$$
Parameters:	T – inter-arrival time at peak cell rate
	Δ – cell slot time
	D – inter-arrival time at peak cell rate, in units of the cell slot time
	τ – cell delay variation tolerance
	MBS – maximum burst size allowed through by leaky bucket
	X – buffer capacity of output port downstream in network
	ρ – admissible load assuming worst-case arrival pattern
Location:	Chapter 11, page 179

Model:	**worst-case analysis of dual leaky bucket (leaky cup and saucer)**
Use:	allowance for impact on load of dual leaky bucket for VBR traffic using SBR transfer capability in ATM

Formulas:

$$MBS = 1 + \left\lfloor \frac{\tau_{IBT}}{T_{SCR} - T_{PCR}} \right\rfloor$$

$$\rho = \frac{2 \cdot \left(\dfrac{X}{MBS} + D \right)}{D \cdot \left(2 - \dfrac{MBS \cdot \ln(CLP)}{X} \right)}$$

Parameters:

T_{PCR} – inter-arrival time at peak cell rate

T_{SCR} – inter-arrival time at sustainable cell rate (SCR)

τ_{IBT} – intrinsic burst tolerance

D – inter-arrival time at sustainable cell rate, in units of the cell slot time

MBS – maximum burst size allowed through by leaky bucket for SCR

X – buffer capacity of output port downstream in network

ρ – admissible load assuming worst-case arrival pattern of maximum-sized bursts arriving at cell slot rate

Location: Chapter 11, page 183

Model:	**multiple ON–OFF sources – excess-rate analysis**
Use:	token bucket configuration for traffic conditioning of aggregate IP flows

Formulas: see multiple ON–OFF sources – excess-rate analysis summarized in Chapter 2, page 27

example relationships between R and B, given different loss probability estimates, are shown in Figure 15.7

Parameters: R – token bucket rate allocation is the same as C, service rate of buffer

B – token bucket capacity is the same as x, the buffer size

both R and B would normally be scaled from packets to octets

Location: Chapter 15, page 266

Dimensioning and configuration

On the longer time scale, the provisioning, dimensioning and configuration of network resources (transmission links, buffer partitions, switching capacities, etc.) are needed to match the expected user demand. This too makes use of queueing analysis to determine probability of blocking or congestion at the connection, or aggregate level. Results for buffer and link dimensioning are summarized below.

Model:	**M/D/1 delay and loss analysis**
Use:	dimensioning small (virtual) buffers for delay-sensitive traffic in IP or ATM
Formulas:	$t_q = s + \dfrac{\rho \cdot s}{2 \cdot (1 - \rho)}$
	see batch arrivals, deterministic service, infinite buffer capacity, summarized in Chapter 2, page 19
	Table 12.2 shows exact buffer dimensions, mean and maximum delay, given offered load and loss probability requirement
Parameters:	s – service time for a cell or packet
	ρ – maximum offered load expected
	t_q – mean delay
Location:	Chapter 12, page 193

Model:	**burst-scale delay and loss analysis**
Use:	dimensioning large (virtual) buffers for delay-insensitive traffic in IP or ATM
Formulas:	$\dfrac{X}{b} = -\dfrac{4 \cdot \rho + 1}{(1 - \rho)^3} \cdot \dfrac{\ln\left(\dfrac{CLP_{target}}{CLP_{bsl}}\right)}{N_0}$
Parameters:	X – buffer capacity in cells or packets
	b – mean number of cells/packets per burst
	ρ – maximum offered load expected
	N_0 – minimum number of active sources for burst-scale queueing (calculated from service rate of buffer, C, and maximum arrival rate, h, for an individual flow or connection)
	CLP_{target} – overall loss probability as design target
	CLP_{bsl} – loss probability contribution from burst-scale loss analysis
Location:	Chapter 12, page 198

Model:	**Erlang's loss analysis**
Use:	dimensioning maximum number of simultaneous flows or connections in IP or ATM, for blocking probability requirement

(continued)

Model:	**Erlang's loss analysis**

Formulas:	$A = a \cdot h$
	$$B = \frac{A^N}{N!} \Bigg/ \left(1 + \frac{A^1}{1!} + \frac{A^2}{2!} + \cdots + \frac{A^N}{N!} \right)$$
	see Table 12.4 for full erlang traffic table
	link dimension is then
	$C = \text{ECR} \cdot N$
Parameters:	h – mean duration of flow or connection
	a – expected arrival rate of flows or connections
	A – offered traffic in erlangs
	B – grade of service (blocking probability) required
	N – number of 'equivalent' circuits
	ECR – equivalent cell rate, i.e. the cell or packet rate required per 'equivalent' circuit (from queueing analysis suitable for traffic type)
	C – total service rate required for (virtual) buffer
Location:	Chapter 3, page 52

3 Teletraffic Engineering

the economic and service arguments

SHARING RESOURCES

A simple answer to the question 'Why have a network?' is 'To communicate information between people'. A slightly more detailed answer would be: 'To communicate information between *all people who would want to exchange information, when they want to*'. Teletraffic engineering addresses the problems caused by sharing of network resources among the population of users; it is used to answer questions like: 'How much traffic needs to be handled?' 'What level of performance should be maintained?' 'What type of, and how many, resources are required?' 'How should the resources be organized to handle traffic?'

MESH AND STAR NETWORKS

Consider a very simple example: a telephone network in which a separate path (with a handset on each end) is provided between every pair of users. For N users, this means having $N(N-1)/2$ paths and $N(N-1)$ telephone handsets. A simple cost-saving measure would be to replace the $N-1$ handsets per user with just one handset and a 1 to $N-1$ switch (Figure 3.1). A total of N handsets and N switches is required, along with the $N(N-1)/2$ paths. If all N users are communicating over the network at the same time, i.e. there are $N/2$ simultaneous calls (or $(N-1)/2$ if N is odd), then $1/(N-1)$ of the paths and all of the handsets and switches would be in use. So in a network with 120 users, for example, the maximum path utilization is just under 1%, and handset and switch utilization are both 100%.

Contrast this with a star network, where each user has a single handset connected to two N to 1 switches, and the poles of the switches are connected by a single path (Figure 3.2). In this example, there are N handsets, $N + 1$ paths, and 2 switches. However, only 2 users may communicate at any one time, i.e. $3/(N + 1)$ of the paths, $2/N$ of the handsets and both of the switches would be in use. So for a network with 120 users, the maximum values are: path utilization is just under 3%, handset utilization is just under 2% and switch utilization is 100%.

In the course of one day, suppose that each one of the 120 users initiates on average two 3-minute calls. Thus the total *traffic volume* is $120 \times 2 \times 3 = 720$ call minutes, i.e. 12 hours of calls. Both star and mesh networks can handle this amount of traffic; the mesh network can carry up to 60 calls simultaneously; the star network carries only 1 call at a time. The mesh network provides the maximum capability for immediate communication, but at the expense of many paths and switches. The star network provides the minimum capability for communication between any two users at minimum cost, but at the inconvenience of having to wait to use the network.

The capacity of the star network could be increased by installing M switching 'units', where each unit comprises two N to 1 switches linked by a single path (Figure 3.3). Thus, with $N/2$ switching units, the star network would have the same communication capability as the mesh network, with the same number of switches and handsets, but requiring only $3N/2$ paths. Even in this case, though, the size becomes impractical as

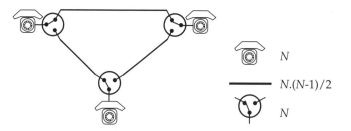

Figure 3.1. The Mesh Network

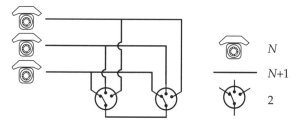

Figure 3.2. The Star Network

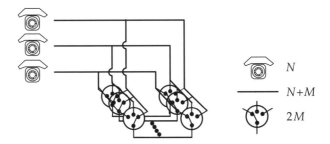

Figure 3.3. The Star Network with M Switching 'Units'

N increases, such that reorganization and further sharing of the switching capacity becomes necessary.

TRAFFIC INTENSITY

Traffic volume is defined as the *total* call holding time for all calls, i.e. the number of calls multiplied by the mean holding time per call. This is not very helpful in determining the total number of paths or switching units required. We need a measure that gives some indication of the *average workload* we are applying to the network.

Traffic intensity is defined in two ways, depending on whether we are concerned with the workload applied to the network (offered traffic), or the work done by the network (carried traffic). The offered traffic intensity is defined as:

$$A = \frac{c \cdot h}{T}$$

where c is number of call attempts in time period T, and h is the mean call holding time (the average call duration). Note that if we let T equal h then the offered traffic intensity is just the number of call attempts during the mean call holding time. The rate of call attempts, also called the 'call arrival rate', is given by

$$a = \frac{c}{T}$$

So the offered traffic intensity can also be expressed as

$$A = a \cdot h$$

For any specific pattern of call attempts, there may be insufficient paths to satisfy all of the call attempts; this is particularly obvious in the case of the star network in Figure 3.2 which has just one path available. A call attempt made when the network is full is blocked (lost) and cannot be carried. If, during time period T, c_c calls are carried and c_l calls are lost, then the total number of call attempts is

$$c = c_c + c_l$$

We then have

$$A = \frac{(c_c + c_l) \cdot h}{T} = C + L$$

where C, the carried traffic, is given by

$$C = \frac{c_c \cdot h}{T}$$

and L, the lost traffic, is given by

$$L = \frac{c_l \cdot h}{T}$$

The blocked calls contributing to the lost traffic intensity obviously do not last for any length of time. The lost traffic intensity, as defined, is thus a theoretical intensity which would exist if there were infinite resources available. Hence the lost traffic cannot be measured, although the *number* of lost calls can. The carried traffic intensity can be measured, and is the average number of paths in use simultaneously (this is intuitive, as we have already stated that it should be a measure of the work being done by the network). As theoretical concepts, however, we shall see that offered, lost and carried traffic prove to be very useful indeed.

Traffic intensity is a dimensionless quantity. It is given the 'honorary' dimension of erlangs in memory of Anders K. Erlang, the founder of traffic theory: one erlang of traffic is written as 1 E. Let's put some numbers in the formulas. In our previous example we had 240 calls over the period of a day, and an average call duration of 3 minutes. Suppose 24 calls are unsuccessful, then $c = 240$, $c_c = 216$, and $c_l = 24$. Thus

$$A = \frac{240 \times 3}{24 \times 60} = 0.5 \, \text{E}$$

$$L = \frac{24 \times 3}{24 \times 60} = 0.05 \, \text{E}$$

$$C = \frac{216 \times 3}{24 \times 60} = 0.45 \, \text{E}$$

Later in this chapter we will introduce a formula which relates A and L according to the number of available paths, N.

It is important to keep in mind that one erlang (1 E) implicitly represents a quantity of bandwidth, e.g. a 64 kbit/s circuit, being used continuously. For circuit-switched telephone networks, it is unnecessary to make this explicit: one telephone call occupies one circuit for the duration of one call. However, if we need to handle traffic with many different bandwidth demands, traffic intensity is rather more difficult to define.

One way of taking the service bandwidth into account is to use the MbitE/s (the 'megabit-erlang-per-second') as a measure of traffic intensity. Thus 1 E of 64 kbit/s digital telephony is represented as 0.064 MbitE/s (in each direction of communication). We shall see later, though, that finding a single value for the service bandwidth of variable-rate traffic is not an easy matter. Suffice to say that we need to know the call arrival rate and the average call duration to give the traffic flow in erlangs, and also the fact that some bandwidth is implicitly associated with the traffic flow for each different type of traffic.

PERFORMANCE

The two different network structures, mesh and star, illustrate how the same volume of traffic can be handled very differently. With the star network, users may have to wait significantly longer for service (which, in a circuit-switched network, can mean repeated attempts by a user to establish a call). A comparison of the waiting time and the delay that users will tolerate (before they give up and become customers of a competing network operator) enables us to assess the adequacy of the network. The waiting time is a measure of performance, as is the 'loss' of a customer.

This also shows a general principle about the flow of traffic:introducing delay reduces the flow, and a reduced traffic flow requires fewer resources. The challenge is to find an optimum value of the delay introduced in order to balance the traffic demand, the performance requirements, and the amount (and cost) of network resources. We will see that much teletraffic engineering is concerned with assessing the traffic flow of cells or packets being carried through the delaying mechanism of the buffer.

TCP: TRAFFIC, CAPACITY AND PERFORMANCE

So we have identified three elements: the capacity of a network and its constituent parts; the amount of traffic to be carried on the network; and the requirements associated with that traffic, in terms of the performance offered by the network to users (see Figure 3.4). One of these elements

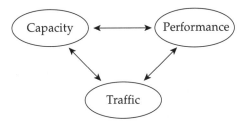

Figure 3.4. Traffic, Capacity and Performance

may be fixed in order to determine how the others vary with each other, or two elements may be fixed in order to find a value for the third. For example, the emphasis in dimensioning is on determining the capacity required, given specific traffic demand and performance targets. Performance engineering aims at assessing the feasibility of a particular network design (or, more commonly, an aspect or part of a network) under different traffic conditions; hence the emphasis is on varying the traffic and measuring the performance for a given capacity (network design). Admission control procedures for calls in an ATM network have the capacity and performance requirements fixed, with the aim of assessing how much, and what mix of, traffic can be accepted by the network.

In summary, a network provides the ability to communicate information between users, with the aim of providing an effective service at reasonable cost. It is uneconomic to provide separate paths between every pair of users. There is thus a need to share paths, and provide users with the means to access these paths when required. A network comprises building blocks (switches, terminal equipment, transmission paths), each of which has a finite capacity for transferring information. Whether or not this capacity is adequate depends on the demand from users for transferring information, and the requirements that users place on that transfer. Teletraffic engineering is concerned with the relationships between these three elements of traffic, capacity and performance.

VARIATION OF TRAFFIC INTENSITY

It is important not to fall into the trap of thinking that a traffic intensity of x erlangs can always be carried on x circuits. The occurrence of any particular pattern of calls is a matter of chance, and the traffic intensity measures the average, not the variation in, traffic during a particular period. The general principle is that more circuits will be needed on a route than the numerical value of the traffic intensity.

Figure 3.5 shows a typical distribution of the number of call attempts per unit time (including the Mathcad code to generate the graph). If we let this 'unit time' be equal to the average call duration, then the average number of 'call attempts per unit time' is numerically equal to the offered traffic intensity. In the case shown it is 2.5 E.

At this stage, don't worry about the specific formula for the Poisson-distribution. The key point is that this distribution is an example which describes the time-varying nature of traffic for a constant average intensity. We could define this average, as before, over the period of one day. But is this sensible? What if 240 calls occur during a day, but 200 of the 240 calls occur between 10 a.m. and 11 a.m.? Then the offered traffic

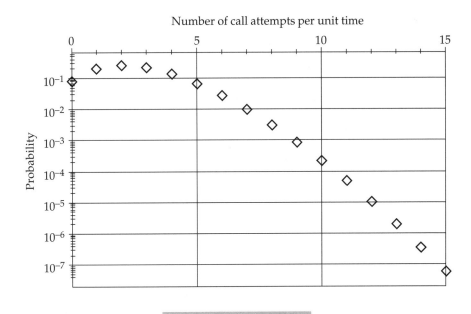

Figure 3.5. Graph of the Distribution of Demand for an Offered Traffic Intensity of 2.5 E, and the Mathcad Code to Generate (x, y) Values for Plotting the Graph

intensity for this hour is

$$A = \frac{200 \cdot 3}{60} = 10 \text{ E}$$

This is significantly larger than the daily average, which we calculated earlier to be 0.5 E. The larger figure for offered traffic gives a better indication of the number of circuits needed. This is because traffic intensity, in practice, varies from a low level during the night to one or more peaks during the day, and a network operator must provide enough circuits to ensure that the performance requirements are met when the traffic is at its peak during the busiest period of the day. The busy hour is defined as a period when the intensity is at a maximum over an uninterrupted period of 60 minutes. Note that the busy-hour traffic is still an average: it is an average over the time scale of the busy hour (recall that this is then the maximum over the time scale of a day).

ERLANG'S LOST CALL FORMULA

In 1917, Erlang published a teletraffic dimensioning method for circuit-switched networks. He developed a formula which expressed the probability, B, of a call being blocked, as a function of the applied (offered) traffic intensity, A, and the number of circuits available, N:

$$B = \frac{A^N}{N!} \bigg/ \left(1 + \frac{A^1}{1!} + \frac{A^2}{2!} + \cdots + \frac{A^N}{N!}\right)$$

B is also the proportion of offered traffic that is lost. Hence

$$B = \frac{L}{A}$$

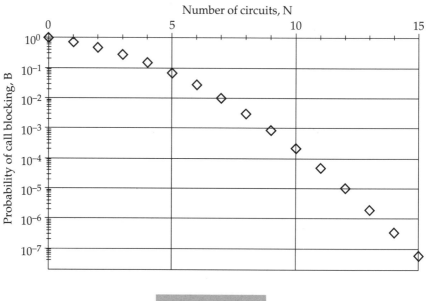

Figure 3.6. Graph of the Probability of Call Blocking for $A = 2.5$ E, and the Mathcad Code to Generate (x, y) Values for Plotting the Graph

where L is the lost traffic, as before. A derivation of Erlang's formula can be found in [3.1, 3.2].

The most important assumption made concerns the pattern of arrivals – calls occur 'individually and collectively at random'. This means they are as likely to occur at one time as at any other time. This type of arrival process is called 'Poisson traffic'. The Poisson distribution gives the probability that a certain number of calls arrive during a particular time interval. We will look at this distribution in more detail in Chapter 6. First, let us plot B against N when $A = 2.5$ E. This is shown in Figure 3.6, with the Mathcad code used to generate the results.

We can read from the graph that the blocking probability is $B = 0.01$ when the number of circuits is $N = 7$. Thus we can use this graph for dimensioning: choose the required probability of blocking and find the number of circuits corresponding to this on the graph. But we don't want to have to produce graphs for every possible value of offered traffic.

TRAFFIC TABLES

The problem is that Erlang's lost call formula gives the call blocking (i.e. loss) probability, B, given a certain number, N, of trunks being offered a certain amount, A, of traffic. But the dimensioning question comes the other way around: with a certain amount, A, of traffic offered, how many trunks, N, are required to give a blocking probability of B? It is not possible to express N in terms of B, so traffic tables, like the one in Table 3.1, have been produced (using iteration), and are widely used, to simplify this calculation.

Table 3.1. Table of Traffic which May Be Offered, Based on Erlang's Lost Call Formula

Number of trunks, N	Probability of blocking, B			
	0.02	0.01	0.005	0.001
	offered traffic, A:			
1	0.02	0.01	0.005	0.001
2	0.22	0.15	0.105	0.046
3	0.60	0.45	0.35	0.19
4	1.1	0.9	0.7	0.44
5	1.7	1.4	1.1	0.8
6	2.3	1.9	1.6	1.1
7	2.9	2.5	2.2	1.6
8	3.6	3.1	2.7	2.1
9	4.3	3.8	3.3	2.6
10	5.1	4.5	4.0	3.1

The blocking probability specified is used to select the correct column, and then we track down the column to a row whose value is equal to or just exceeds the required offered traffic intensity. The value of N for this row is the minimum number of circuits needed to satisfy the required demand at the specified probability of call blocking. From the columns of data in the traffic table, it can be seen that as the number of circuits increases, the average loading of each circuit increases, for a fixed call-blocking probability. This is plotted in Figure 3.7 (note that for simplicity we approximate the circuit loading by the average *offered* traffic per circuit, A/N). So, for example, if we have 10 circuits arranged into two groups of 5, then for a blocking probability of 0.001 we can load each group with 0.8 E, i.e. a total of 1.6 E. If all 10 circuits are put together into one group, then 3.1 E can be offered for the same probability of blocking of 0.001. In the first case the offered traffic per circuit is 0.16 E; in the second it is 0.31 E. Thus the larger the group of circuits, the better the circuit loading efficiency.

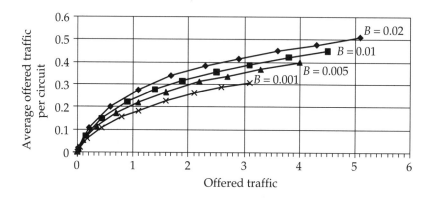

Figure 3.7.   Loading Efficiency of Circuits

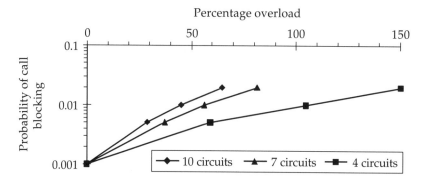

Figure 3.8.   Overload Capability of Circuit Groups

However, if we consider how a group of circuits perform under overload, there are disadvantages in having large groups. Here, we use the rows of data from Table 3.1 and plot, in Figure 3.8, the blocking probability against the percentage increase in offered traffic over the offered traffic for $B = 0.001$. Small groups of circuits do better under overload conditions than larger groups; this is because the inefficient small groups have more 'waste' capacity to deal with unexpected overload and the deterioration in the blocking probability is small. For a large group of circuits this deterioration can be substantial.

4 Performance Evaluation

how's it going?

METHODS OF PERFORMANCE EVALUATION

If we are to design a network, we need to know whether the equipment is going to be used to best effect, and to achieve this we will need to be able to evaluate its performance. Methods for performance evaluation fall into two categories: measurement techniques, and predictive techniques; with the latter category comprising mathematical analysis and simulation.

Measurement

Measurement methods require real networks to be available for experimentation. The advantage of direct measurement of network performance is that no detail of network operation is excluded: the actual operation of the real network is being monitored and measured. However, there are some constraints. A revenue-earning network cannot be exercised to its limits of performance because customers are likely to complain and take their business elsewhere. An experimental network may be limited in the number and type of traffic sources available, thus restricting the range of realistic experimental conditions.

Predictive evaluation: analysis/simulation

In comparing analysis and simulation, the main factors to consider are the accuracy of results, the time to produce results, and the overall cost of using the method (this includes development as well as use).

One advantage of analytical solutions is that they can be used reasonably quickly. However, the need to be able to solve the model restricts the range of system or traffic characteristics that can be included. This can result in right answers to the wrong problem, if the model has to be changed so much from reality to make it tractable. Thus analysis is often used to produce an approximation of a system, with results being produced relatively quickly and cheaply.

Networks of almost arbitrary complexity can be investigated using simulation: systems may be modelled to the required level of precision. Very often, simulation is the only feasible method because of the nature of the problem and because analytical techniques become too difficult to handle. However, simulation can be costly to develop and run, and it is time-consuming, particularly when very rare events (such as ATM cell loss) are being measured (although accelerated simulation techniques can reduce the time and cost involved).

QUEUEING THEORY

Analysis of the queueing process is a fundamental part of performance evaluation, because queues (or 'waiting lines') form in telecommunications systems whenever customers contend for limited resources. In technologies such as ATM or IP not only do connections contest, and may be made to queue, but each accepted connection consists of a stream of cells or packets and these also must queue at the switching nodes or routers as they traverse the network.

We will use a queue then as a mathematical expression of the idea of resource contention (Figure 4.1): customers arrive at a queueing system needing a certain amount of service; they wait for service, if it is not immediately available, in a storage area (called a 'buffer', 'queue', or 'waiting line'); and having waited a certain length of time, they are served and leave the system. Note that the term 'customers' is the general expression you will encounter in queueing theory terminology and it is

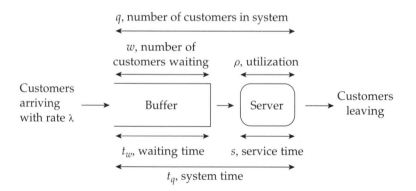

Figure 4.1. The Queueing System

used to mean 'anything that queues'; in ATM or IP, the customers can be cells, packets, bursts, flows, or connections. In the rest of this chapter, the queueing systems refer to ATM buffers and the customers are cells.

Any queueing system is described by the arrival pattern of customers, the service pattern of customers, the number of service channels, and the system capacity. The arrival pattern of customers is the input to a queueing system and can sometimes be specified just as the average number of arrivals per unit of time (mean arrival rate, λ) or by the average time between arrivals (mean inter-arrival time). The simplest input any queueing system can have is 'deterministic', in which the arrival pattern is one customer every t time units, i.e. an arrival rate of $1/t$. So, for a 64 kbit/s constant bit-rate (CBR) service, if all 48 octets of the information field are filled then the cell rate is 167 cell/s, and the inter-arrival time is 6 ms. If the arrival pattern is 'stochastic' (i.e. it varies in some random fashion over time), then further characterization is required, e.g. the probability distribution of the time between arrivals. Arrivals may come in batches instead of singly, and the size of these batches may vary. We will look at a selection of arrival patterns in Chapter 6.

The service pattern of customers, as with arrival patterns, can be described as either a rate, μ, of serving customers, or as the time, s, required to service a customer. There is one important difference: service time or service rate are conditioned on the system not being empty. If it is empty, the service facility is said to be 'idle'. However, when an ATM cell buffer is empty, a continuous stream of empty cell slots is transmitted. Thus the server is synchronized and deterministic; this is illustrated in Figure 1.4

In the mathematical analysis of an ATM buffer, the synchronization is often neglected – thus a cell is assumed to enter service immediately upon entry to an empty buffer, instead of waiting until the beginning of the next free slot. For a 155.52 Mbit/s link, the cell slot rate is 366 792 cell/s and the service time per cell is 2.726 μs. However, 1 in every 27 cell slots is used for operations and maintenance (OAM) cells for various monitoring and measurement duties. Thus the cell slot rate available for traffic is

$$\frac{26}{27} \cdot 366\,792 = 353\,208 \text{ cell/s}$$

which can be approximated as a service time per cell of 2.831 μs.

The number of service channels refers to the number of servers that can serve customers simultaneously. Multi-channel systems may differ according to the organization of the queue(s): each server may have its own queue, or there may be only one queue for all the servers. This is of particular interest when analysing different ATM switch designs.

The system capacity consists of the waiting area and the number of service channels, and may be finite or infinite. Obviously in a real system the capacity must be finite. However, assuming infinite capacity can simplify the analysis and still be of value in describing ATM queueing behaviour.

Notation

Kendall's notation, $A/B/X/Y/Z$, is widely used to describe queueing systems:

A specifies the inter-arrival time distribution
B specifies the service time distribution
X specifies the number of service channels
Y specifies the system capacity, and
Z specifies the queue discipline

An example is the M/D/1 queue. Here the 'M' refers to a memoryless, or Markov, process, i.e. negative exponential inter-arrival times. The 'D' means that the service time is always the same: fixed or 'deterministic' (hence the D), and '1' refers to a single server. The Y/Z part of the notation is omitted when the system capacity is infinite and the queue discipline is first-come first-served. We will introduce abbreviations for other arrival and service processes as we need them.

Elementary relationships

Table 4.1 summarizes the notation commonly used for the various elements of a queueing process. This notation is not standardized, so beware... for example, q may be used, either to mean the average number of customers in the system, or the average number waiting to be served (unless otherwise stated, we will use it to mean the average number in the system).

There are some basic queueing relationships which are true, assuming that the system capacity is *infinite*, but regardless of the arrival or service

Table 4.1. Commonly Used Notation for Queueing Systems

Notation	Description
λ	mean number of arrivals per unit time
s	mean service time for each customer
ρ	utilization; fraction of time the server is busy
q	mean number of customers in the system (waiting or being served)
t_q	mean time a customer spends in the system
w	mean number of customers waiting to be served
t_w	mean time a customer spends waiting for service

patterns and the number of channels or the queue discipline. The utilization, ρ, is equal to the product of the mean arrival rate and the mean service time, i.e.

$$\rho = \lambda \cdot s$$

for a single-server queue. With one thousand 64 kbit/s CBR sources, the arrival rate is 166 667 cell/s. We have calculated that the service time of a cell is 2.831 µs, so the utilization, ρ, is 0.472.

The mean number of customers in the queue is related to the average time spent waiting in the queue by a formula called Little's formula (often written as $L = \lambda \cdot W$). In our notation this is:

$$w = \lambda \cdot t_w$$

So, if the mean waiting time is 50 µs, then the average queue length is 8.333 cells. This relationship also applies to the average number of customers in the system:

$$q = \lambda \cdot t_q$$

The mean time in the system is simply equal to the sum of the mean service time and waiting time, i.e.

$$t_q = t_w + s$$

which, in our example, gives a value of 52.831 µs. The mean number of customers in a single-server system is given by

$$q = w + \rho$$

which gives a value of 8.805 cells.

The M/M/1 queue

We can continue with the example of N CBR sources feeding an ATM buffer by making two assumptions, but the example will at least give us a context for choosing various parameter values. The first assumption is that the cell arrival pattern from N CBR sources can be approximated by negative exponential inter-arrival times. This is the same as saying that the arrivals are described by a Poisson process. This process just looks at the arrival pattern from a different perspective. Instead of specifying a time duration, the Poisson distribution counts the number of arrivals in a time interval.

The second assumption is that the service times of these cells are described by a negative exponential distribution. In Chapter 8 we will see that the first assumption can be justified for large N. Given the fact that ATM uses fixed-length cells (and hence fixed service times), the second assumption is not very accurate! Nonetheless, we can use this example to illustrate some important points about queueing systems.

So, how large should we make the ATM buffer? Remember that the M/M/1 queueing system assumes infinite buffer space, but we can get some idea by considering the average number of cells in the system, which is given by

$$q = \frac{\rho}{1 - \rho}$$

In our example, the utilization resulting from 1000 CBR sources is 0.472, which gives an average system size of 0.894 cell. Subtracting the utilization from this gives us the average waiting space that is used, 0.422 cell. This is not a very helpful result for dimensioning an ATM buffer; we would expect to provide at least some waiting space in excess of 1 cell. But if we look at a graph (Figure 4.2) of q against ρ, as ρ varies from 0 to 1, then we can draw a very useful conclusion. The key characteristic is the 'knee' in the curve around 80% to 90% utilization,

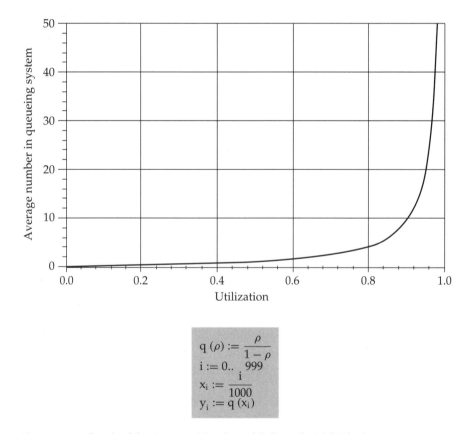

Figure 4.2. Graph of the Average Number of Cells in the M/M/1 Queueing System, and the Mathcad Code to Generate (x, y) Values for Plotting the Graph

which suggests that it is best to operate the system below 80% utilization to avoid large queues building up.

But we still do not have any idea of how large to make the ATM buffer. The next step is to look at the distribution of system size which is given by

$$\Pr\{\text{system size} = x\} = (1 - \rho)\rho^x$$

Figure 4.3 shows this distribution for a range of different utilization values, including the value of 0.472 which is our particular example. In this case we can read from the graph that the probability associated with a system size of 10 cells is 0.0003.

From this we might conclude that a buffer length of 10 cells would *not* be adequate to meet the cell loss probability (CLP) requirements of ATM which are often quoted as being 10^{-8} or less. For the system size probability to be less than 10^{-8}, the system size needs to be 24 cells; the actual probability is 7.89×10^{-9}. In making this deduction, we have approximated the CLP by the probability that the buffer has reached a

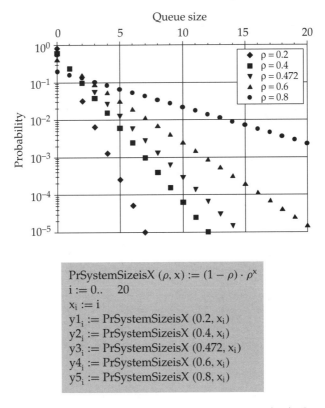

Figure 4.3. Graph of the System State Distribution for the M/M/1 Queue, and the Mathcad Code to Generate (x, y) Values for Plotting the Graph

particular level in our infinite buffer model. This assumes that an infinite buffer model is a good model of a finite buffer, and that Pr{system size = x} is a reasonable approximation to the loss from a finite queue of size x.

Before we leave the M/M/1, let's look at another approximation to the CLP. This is the probability that the system size *exceeds* x. This is found by summing the state probabilities up to and including that for x, and then subtracting this sum from 1 (this is a simpler task than summing from $x + 1$ up to infinity). The equation for this turns out to be very simple:

$$\text{Pr}\{\text{system size} > x\} = \rho^{x+1}$$

When $x = 24$ and $\rho = 0.472$, this equation gives a value of 7.06×10^{-9} which is very close to the previous estimate.

Now Figure 4.4 compares the results for the two approximations, Pr{system size = x} and Pr{system size $\geqslant x$}, with the actual loss probability from the M/M/1/K system, for a system size of 24 cells, with the utilization varying from 0 to 1. What we find is that all three approaches give very similar results over most utilization values, diverging only when the utilization approaches 100%. For the example utilization value of 0.472, there is in fact very little difference. The main point to note here is that an infinite queue can provide a useful approximation for a finite one.

The M/D/1/K queue

So let's now modify our second assumption, about service times, and instead of being described by a negative exponential distribution we will model the cells as they are – of fixed length. The only assumption we will make now is that they enter service whenever the server is idle, rather than waiting for the next cell slot. The first assumption, about

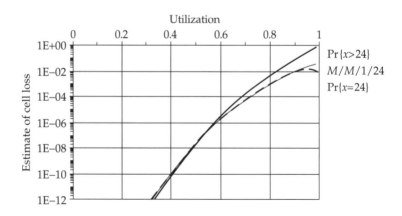

Figure 4.4. Comparison of CLP Estimates for Finite M/M/1 Queueing System

arrival times, remains the same. We will deal with a finite queue directly, rather than approximating it to an infinite queue. This, then, is called the M/D/1/K queueing system.

The solution for this system is described in Chapter 7. Figure 4.5 compares the cell loss from the M/D/1/K with the M/M/1 CLP estimator, Pr{system size = x}, when the system size is 10. As before, the utilization ranges from 0 to 1. At the utilization of interest, 0.472, the difference between the cell loss results is about two orders of magnitude.

So we need to remember that performance evaluation 'answers' can be rather sensitive to the choice of model, and that this means they will always be, to some extent, open to debate. For the cell loss probability in the M/D/1/K to be less than 10^{-8}, the system size needs to be a minimum of 15 cells, and the actual CLP (if it is 15 cells) is 4.34×10^{-9}. So, by using a more accurate model of the system (compared to the M/M/1), we can save on designed buffer space, or alternatively, if we use a system size of 24 cells, the utilization can be increased to 66.8%, rather than 47.2%. This increase corresponds to 415 extra 64 kbit/s simultaneous CBR connections.

It is also worth noting from Figure 4.5 that the cell loss probabilities are very close for high utilizations, i.e. the difference between the two models, with their very different service time assumptions, becomes almost negligible under heavy traffic conditions. In later chapters we present some useful heavy traffic results which can be used for performance evaluation of ATM, where applicable.

Delay in the M/M/1 and M/D/1 queueing systems

ATM features both cell loss and cell delay as key performance measures, and so far we have only considered loss. However, delay is particularly

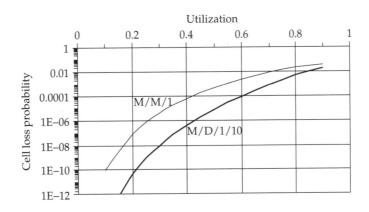

Figure 4.5.   Comparison of M/D/1/K and M/M/1 Cell Loss Results

important to real-time services, e.g. voice and video. Little's result allows us to calculate the average waiting time from the average number waiting in the queue and the arrival rate. If we apply this analysis to the example of 1000 CBR connections multiplexed together, we obtain the following:

$$t_w = \frac{w}{\lambda} = \frac{0.422}{166\,667} = 2.532 \ \mu s$$

The average time in the system is then

$$t_q = t_w + s = 2.532 + 2.831 = 5.363 \ \mu s$$

Another way of obtaining the same result is to use the waiting time formula for the M/M/1 queue. This is

$$t_w = \frac{\rho \cdot s}{1 - \rho}$$

For the M/D/1 queue, there is a similar waiting time formula:

$$t_w = \frac{\rho \cdot s}{2 \cdot (1 - \rho)}$$

In both cases we need to add the service time (cell transmission time) to obtain the overall delay through the system. But the main point to note is that the average waiting time in the M/D/1 queue (which works out as 1.265 μs in our example) is *half* that for the M/M/1 queue.

Figure 4.6 shows the average waiting time against utilization for both queue models. The straight line shows the cell service time. Notice how it dominates the delay up to about 60% utilization. We can take as a useful 'rule of thumb' that the average delay arising from queueing across a network will be approximately twice the sum of the service times. This assumes, of course, that the utilization in any queue will be no more than about 60%. For the total end-to-end delay, we must also add in the propagation times on the transmission links.

So, are these significant values? Well, yes, but, taken alone, they are not sufficient. We should remember that they are averages, and cells will actually experience delays both larger and smaller. Delay is particularly important when we consider the end-to-end characteristics of connections; all the cells in a connection will have to pass through a series of buffers, each of which will delay them by some 'random' amount depending on the number of cells already in the buffer on arrival. This will result in certain cells being delayed more than others, so-called delay jitter, or cell delay variation (CDV).

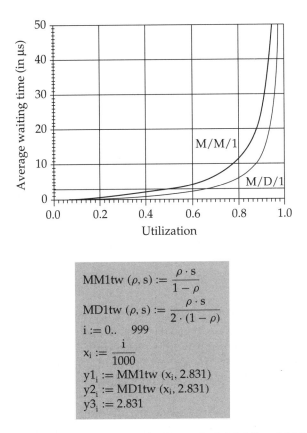

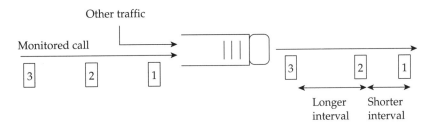

Figure 4.6. Graph of the Average Waiting Times for M/M/1 and M/D/1 Queues, and the Mathcad Code to Generate (x, y) Values for Plotting the Graph

Figure 4.7. Variation in Delay for Cells Passing through a Buffer

A pictorial illustration of this is shown in Figure 4.7. Here, we show only the cells of the connection we are monitoring; there is, of course, other traffic to contribute to the queueing in the buffer. The second cell experiences a shorter delay than the first and third cells. This produces a smaller interval between cells 1 and 2, and a longer interval between cells 2 and 3. Variation in delay can be a particular problem for usage parameter control, and we will look at this issue again in Chapter 11.

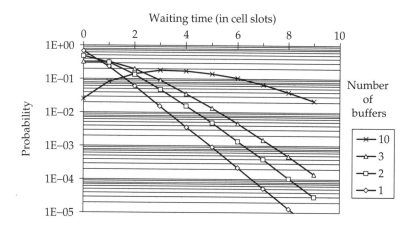

Figure 4.8.   End-to-End Waiting Time Distributions

So much for illustrations, what of concrete examples? If again we use our CBR example (1000 multiplexed CBR 64 kbit/s sources), we can use more of the theory associated with the M/D/1 queue to predict the result of passing this stream of cells through a succession of similar queues, and plot the resulting waiting time distribution. The probabilities associated with a cell in the stream being delayed by x time slots having passed through 1, 2, 3 and 10 similar buffers are shown in Figure 4.8. To generate these results, we have assumed that each buffer is independent of all the others, and that they are all loaded at 0.472. The results clearly show the trend for the delay distribution to flatten as the number of buffers increases: as you might expect, the more buffers the cells pass through, the more the probabilities associated with long waits and with short waits tend to equal out.

5 Fundamentals of Simulation

those vital statistics

DISCRETE TIME SIMULATION

This chapter is intended as an *introduction* to simulation and, in particular, its application to cell- and packet-based queueing. For anyone wanting a more comprehensive treatment of the subject of simulation in general, we refer to [5.1].

We will introduce the subject of simulation by concentrating on a discrete version of the $M/D/1$ queue, applicable to the study of ATM cell buffering. There are two basic ways to simulate such a queue:

- discrete *time* advance
- discrete *event* advance

In the former, the simulator moves from time instant i to time instant $i+1$ regardless of whether the system state has changed, e.g. if the $M/D/1$ queue is empty at i it could still be empty at $i+1$ and the program will still only advance the clock to time $i+1$. These instants can correspond to cell slots in ATM. In discrete-event simulation, the simulator clock is advanced to the next time for which there is a change in the state of the simulation model, e.g. a cell arrival or departure at the $M/D/1$ queue.

So we have a choice: discrete time advance or discrete event advance. The latter can run more quickly because it will cut out the slot-to-slot transitions when the queue is empty, but the former is easier to understand in the context of ATM because it is simpler to implement and it models the cell buffer from the point of view of the server process, i.e. the 'conveyor belt' of cell slots (see Figure 1.4). We will concentrate on the discrete time advance mechanism in this introduction.

In the case of the synchronized M/D/1 queue the obvious events between which the simulator can jump are the end of time slot instants, and so the simulator needs to model the following algorithm:

$$K_i = \max(0, K_{i-1} + A_i - 1)$$

where

K_i = number of cells in modelled system at end of time slot i
A_i = number of cells arriving to the system during time slot i

This algorithm can be expressed as a simulation program in the following pseudocode:

```
BEGIN
  initialize variables
    i, A, K, arrival rate, time slot limit, histogram[]
  WHILE (i < time slot limit)
  generate new arrivals
    A := Poisson(arrival rate)
    K := K + A
  serve a waiting cell
    IF K > 0 THEN
      K := K - 1
    ELSE
      K := 0
  store results
    histogram[K] := histogram[K] + 1
  advance time to next time slot
    i := i + 1
  END WHILE
END
```

The main program loop implements the discrete time advance mechanism in the form of a loop counter, i. The beginning of the loop corresponds to the start of time slot i, and the first section *'generate new arrivals'* calls function 'Poisson' which returns a random non-negative integer for the number of cell arrivals during this current time slot. We model the queue with an arrivals-first buffer management strategy, so the service instants occur at the end of the time slot after any arrivals. This is dealt with by the second section, *'serve a waiting cell'*, which decrements the queue state variable K, if it is greater than 0, i.e. if the queue is not empty. At this point, in *'store results'* we record the state of the queue in a histogram. This is simply a count of the number of times the queue is in state K, for each possible value of K, (see Figure 5.1), and can be converted to an estimate of the state probability distribution by dividing each value in the array 'histogram[]' by the total number of time slots in the simulation run.

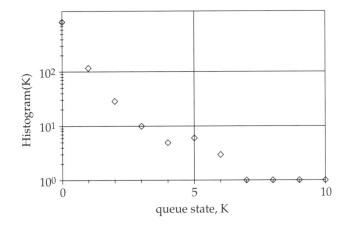

Figure 5.1.   An Example of a Histogram of the Queue State (for a Simulation Run of 1000 Time Slots)

Generating random numbers

The function 'Poisson' generates a random non-negative integer number of cell arrivals according to a Poisson distribution with a particular arrival rate. This is achieved in two parts: generate random numbers that are uniformly distributed over the range 0 to 1; convert these random numbers to be Poisson-distributed. Let's assume that we have a function 'generate random number' which implements the first part. The following pseudocode converts the random numbers from having a uniform distribution to having a Poisson distribution.

```
FUNCTION X = Poisson(arrival rate)
  initialize variables
    a := e^(-arrival rate)
    b := 1
    j := -1
  REPEAT
    j := j + 1
    U := generate random number
    b := b · U
  UNTIL (b < a)
  return result
    X := j
END FUNCTION
```

The REPEAT loop corresponds to the 'generation' of cells, and the loop records the number of cells in the batch in variable j, returning the final total in variable X. Remember that with this particular simulation program we are not interested in the arrival time of each cell within the slot, but in the *number* of arrivals during a slot.

But how do we generate the random numbers themselves? A good random number generator (RNG) should produce a sequence of numbers which are uniformly distributed on the range [0, 1] and which do not exhibit any correlation between the generated numbers. It must be fast and avoid the need for much storage. An important property of the random number sequence is that it must be reproducible; this aids debugging, and can be used to increase the precision of the results.

A typical RNG is of the form:

$$U_i = (a \cdot U_{i-1} + c) \bmod (m)$$

where U_i is the ith random number generated, and m (the modulus), a (the multiplier) and c (the increment) are all non-negative integers, as is U_0, the initial value which is called the 'seed'. The values should satisfy $0 < m, a < m, c < m$ and $U_0 < m$. In practice m is chosen to be very large, say 10^9.

Obviously, once the RNG produces a value for U_i which it has produced before, the sequence of numbers being generated will repeat, and unwanted correlations will begin to appear in the simulator results. An important characteristic of an RNG is the length of the sequence before it repeats; this is called the 'period'. The values of m and c are chosen, in part, to maximize this period. The Wichmann–Hill algorithm combines three of these basic generators to produce a random number generator which exhibits exceptional performance. The pseudocode for this algorithm is:

```
FUNCTION U = generate random number
  x := (171 · x) mod(30269)
  y := (172 · y) mod(30307)
  z := (170 · z) mod(30323)
  U := (x/30269) + (y/30307) + (z/30323)
  temp := trunc(U)
  U := U - temp
END FUNCTION
```

The period is of particular relevance for ATM traffic studies, where rare events can occur with probabilities as low as 10^{-10} (e.g. lost cells). Once an RNG repeats its sequence, unwanted correlations will begin to appear in the results, depending on how the random number sequence has been applied. In our discrete time advance simulation, we are simulating time slot by time slot, where each time slot can have 0 or more cell arrivals. The RNG is called once per time slot, and then once for each cell arrival during the time slot. With the discrete event advance approach, a cell-by-cell simulator would call the RNG once per cell arrival to generate the inter-arrival time to the next cell.

The Wichmann–Hill algorithm has a period of about 7×10^{12}. Thus, so long as the number of units simulated does not exceed the period of 7×10^{12}, this RNG algorithm can be applied. The computing time required to simulate this number of cells is impractical anyway, so we can be confident that this RNG algorithm will not introduce correlation due to repetition of the random number sequence. Note that the period of the Wichmann–Hill algorithm is *significantly* better than many of the random number generators that are supplied in general-purpose programming languages. So, check carefully before you use a built-in RNG.

Note that there are other ways in which correlations can appear in a sequence of random numbers. For more details, see [5.1].

M/D/1 queue simulator in Mathcad

The following Mathcad code implements the discrete time advance simulator pseudocode for the M/D/1 queue. Note that the WHILE loop in the main pseudocode is vectorized (using range variable i), as is the REPEAT loop in the Poisson function pseudocode (using range variable j). An example of the histogram of queue state results is shown in Figure 5.1 (plotting histogramK against Kbins).

initialize variables

time slotlimit := 1000
arrivalrate := 0.5
$i := 1 ..$ time slotlimit
maxK := 10
$K_0 := 0$

generate new arrivals

$a := e^{-arrivalrate}$
$b_{i,0} := 1$
$j := 1 .. 10$
$\overrightarrow{b_{i,j} := (rnd(1) \cdot b_{i,j-1})}$
$cells_{i,j} := if(b_{i,j} < a, 0, 1)$
$A_i := \sum_j cells_{i,j}$

serve a waiting cell

$K_i := max[[0 \quad (K_{i-1} + A_i) - 1]]$

store results

$$\text{actualload} := \frac{\sum_i A_i}{\text{time slotlimit}}$$

$\text{actualload} = 0.495$

$q := 0, 1 ..\ \text{maxK}$

$\text{Kbins}_q := q$

$\text{histogramK} := \text{hist (Kbins , K)}$

end of simulation

Reaching steady state

When do we stop a simulation? This is not a trivial question, and if, for example, we want to find the cell loss probability in an M/D/1/K model, then the probability we are seeking is actually a 'steady-state' probability: the long-run proportion of cells lost during period T as $T \to \infty$. Since we cannot actually wait that long, we must have some prior idea about how long it takes for the simulator to reach a good approximation to steady-state.

A simulation is said to be in steady state, not when the performance measurements become constant, but when the distribution of the measurements becomes (close to being) invariant with time. In particular, the simulation needs to be sufficiently long that the effect of the initial state of the system on the results is negligible. Let's take an example. Recall from Chapter 4 that we can use the probability that the queue size is greater than K, denoted $Q(K)$, as an estimate of the cell loss from a finite queue of size K. Suppose that the queue length is 2. We can calculate $Q(2)$ from the histogram data recorded in our simulation program thus:

$$Q(2) = \frac{\sum_{K=3}^{\infty} \text{histogram}[K]}{i}$$

or, alternatively as

$$Q(2) = \frac{i - \sum_{K=0}^{2} \text{histogram}[K]}{i}$$

If we start our M/D/1 simulator, and plot $Q(2)$ for it as this value evolves over time, we will see something like that which is shown in Figure 5.2.

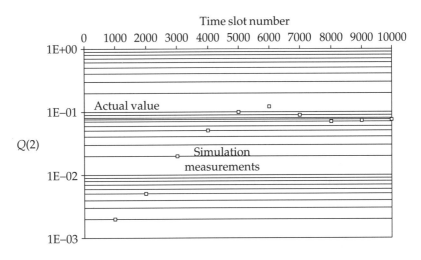

Figure 5.2. Evolution of $Q(2)$ for the Simulated M/D/1

Here, the simulator calculates a measurement result for $Q(2)$ every 1000 time slots; that is to say it provides an *estimate* of $Q(2)$ every 1000 slots. But from Figure 5.2 we can see that there are 'transient' measurements, and that these strongly reflect the initial system state. It is possible to cut out these measurements in the calculation of steady state results; however, it is not easy to identify when the transient phase is finished. We might consider the first 7000 slots as the transient period in our example.

Batch means and confidence intervals

The output from one run of a simulation is a sequence of measurements which depends on the particular sequence of random numbers used. In the example we have been considering, we store results at the end of each time slot, then, at intervals of 1000 time slots, we output a value for $Q(2)$. But we do not take the last value to be output as the final 'result' of the simulation run. The *sequence* of measurements of $Q(2)$ needs to be evaluated statistically in order to provide reliable results for the steady-state value of $Q(2)$.

Suppose that we take $j = 1$ to N measurements of $Q(2)$. First, we can obtain an estimate of the mean value by calculating

$$\hat{Q}(2) = \frac{\sum_{j=1}^{N} Q(2)_j}{N}$$

Then we need an estimate of how the measurements vary over the set. We can construct an estimate of the *confidence interval* for $Q(2)$ by calculating

$$\hat{Q}(2) \pm z_{\alpha/2} \cdot \frac{\sum_{j=1}^{N}(Q(2)_j - \hat{Q}(2))^2}{N \cdot (N-1)}$$

where $z_{\alpha/2}$ is obtained from standard normal tables and $1 - \alpha$ is the degree of confidence.

A confidence interval quantifies the confidence that can be ascribed to the results from a simulation experiment, in a statistical sense. For example, a 90% confidence interval (i.e. $\alpha = 0.1$) means that for 90% of the simulation runs for which an interval is calculated, the actual value for the measure of interest falls within the calculated interval (see Figure 5.3). On the other 10% of occasions, the actual value falls outside the calculated interval. The actual percentage of times that a confidence interval does span the correct value is called the 'coverage'.

There are a number of different methods for organizing simulation experiments so that confidence intervals can be calculated from the measurements. The method of independent replications uses N estimates obtained from N independent simulation runs. In the method of batch means, one single run is divided into N batches (each batch of a certain fixed number, L, of observations) from which N estimates are calculated. The value of L is crucial, because it determines the correlation between batches: considering our M/D/1 example again, if L is too small then the system state at the end of N_j will be heavily influenced by (correlated with) the system state at the end of N_{j-1}. The regenerative method also uses a single run, but depends on the definition of a regenerative state – a state after which the process repeats probabilistically. Determining an appropriate regenerative state can be difficult, and it can be time-consuming to obtain a sufficient number of points at which the simulation passes through such states, in order to calculate valid confidence intervals.

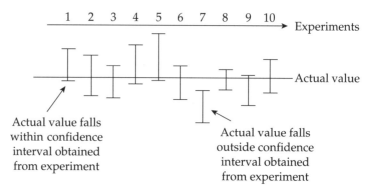

Figure 5.3.   Confidence Intervals and Coverage

The main advantage of methods involving just a single run is that only one transient phase needs to be discarded. Determining the best length for the simulation run(s) is a problem for all three methods. This is because, if the runs are too short, they can produce confidence intervals with actual coverage considerably lower than desired. However, this has to be balanced against the need to limit the sample size to minimize the time (and hence, cost) of the simulation; so the emphasis is on finding a sufficient sample size. In addressing this problem, an alternative to the arbitrary fixed sample size approach is to increase the sample size sequentially until an acceptable confidence interval can be constructed.

Validation

Any simulation model will need to be checked to ensure that it works. This can be a problem: a very general program that is capable of analysing a large number of scenarios will be impossible to test in all of them, especially as it would probably have been developed to solve systems that have no analytical solution to check against. However, even for the most general of simulators it will be possible to test certain simple models that do have analytical solutions, e.g. the $M/D/1$.

ACCELERATED SIMULATION

In the discussion on random number generation we mentioned that the computing time required to simulate 10^{12} cells is impractical, although cell loss probabilities of 10^{-10} are typically specified for ATM buffers. In fact, most published simulation results for ATM extend no further than probabilities of 10^{-5} or so.

How can a simulation be accelerated in order to be able to measure such rare events? There are three main ways to achieve this: use more computing power, particularly in the form of parallel processing; use statistical techniques to make better use of the simulation measurements; and decompose the simulation model into connection, burst and cell scales and use only those time scales that are relevant to the study.

We will focus on the last approach because it extends the analytical understanding of the cell and burst scales that we develop in later chapters and applies it to the process of simulation. In particular, burst-scale queueing behaviour can be modelled by a technique called 'cell-rate simulation'.

Cell-rate simulation

The basic unit of traffic with cell-rate simulation is a 'burst of cells'. This is defined as a fixed cell-rate lasting for a particular time period

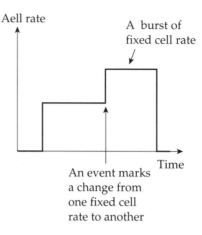

Figure 5.4.   The Basic Unit of Traffic in Cell-Rate Simulation

during which it is assumed that the inter-arrival times do not vary (see Figure 5.4). Thus instead of an event being the arrival or service of a cell, an event marks the change from one fixed cell-rate to another. Hence traffic sources in a cell-rate simulator must produce a sequence of *bursts* of cells. Such traffic sources, based on a cell-rate description are covered in Chapter 6.

The multiplexing of bursts from different sources through an ATM buffer has to take into account the simultaneous nature of these bursts. Bursts from different sources will overlap in time and a change in the rate of just one source can affect the output rates of all the other VCs passing through the buffer.

An ATM buffer is described by two parameters: the maximum number of cells it can hold, i.e. its buffer capacity; and the constant rate at which cells are served, i.e. its cell service-rate. The state of a queue, at any moment in time, is determined by the combination of the input rates of all the VCs, the current size of the queue, and the queue parameter values.

The flow of traffic through a queue is described by input, output, queueing and loss rates (see Figure 5.5). Over any time period, all cells input to the buffer must be accounted for; they are or served queued or lost. At any time, the rates for each VC, and for all VCs, must balance:

$$\text{input rate} = \text{output rate} + \text{queueing rate} + \text{loss rate}$$

When the queue is empty, the output rates of VCs are equal to their input rates, the total input rate is less than the service rate, and so there is no burst-scale queueing.

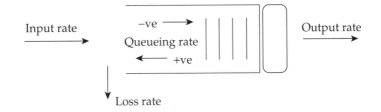

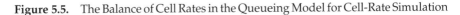

Figure 5.5. The Balance of Cell Rates in the Queueing Model for Cell-Rate Simulation

$$\text{output rate} = \text{input rate}$$

$$\text{queueing rate} = \text{loss rate} = 0$$

In a real ATM system there will of course be cell-scale queueing, but this behaviour is not modelled by cell-rate simulation. When the combined input rate exceeds the service rate, the queue size begins to increase at a rate determined by the difference between the input rate and service rate of the queue.

$$\text{queueing rate} = \text{input rate} - \text{service rate}$$

For an individual VC, its share of the total queueing rate corresponds to its share of the total input rate. Once the queue becomes full, the total queueing rate is zero and the loss rate is equal to the difference in the input rate and service rate.

$$\text{loss rate} = \text{input rate} - \text{service rate}$$

Although this appears to be a simple model for the combined cell rates, it is more complicated when individual VCs are considered. An input change to a full buffer, when the total input rate exceeds the service rate, has an impact not only on the loss rate but also on all the individual VC queueing rates. Also, the effect of a change to the input rate of a VC, i.e. an event at the input to the queue, is not immediately apparent on the output, if there are cells queued. At the time of the input event, only the queueing and/or loss rates change. The change appears on the output only after the cells which are currently in the queue have been served. Then, at the time of this output event, the queueing and output rates change.

It is beyond the scope of this book to describe the cell-rate simulation technique in more detail (the interested reader is referred to [5.2]); however, we present some results in Figure 5.6 which illustrate the accelerated nature of the technique. In comparison with cell-by-cell simulation, cell-rate simulation shows significant speed increases, varying from 10 times to over 10 000 times faster. The speed improvement increases in

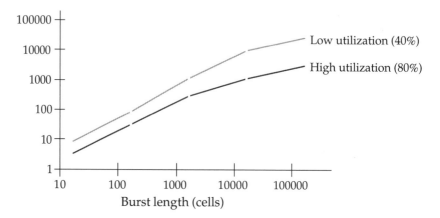

Figure 5.6. Speed Increase of Cell-Rate Simulation Relative to Cell-by-Cell Simulation

proportion to the average number of cells in a fixed-rate burst, and also increases the lower the utilization and hence also the lower the cell loss. This is because it focuses processing effort on the traffic behaviour which dominates the cell loss: the burst-scale queueing behaviour. So cell-rate simulation enables the low cell-loss probabilities required of ATM networks to be measured within reasonable computing times.

6 Traffic Models

you've got a source

LEVELS OF TRAFFIC BEHAVIOUR

So, what kind of traffic behaviour are we interested in for ATM, or IP? In Chapter 3 we looked at the flow of calls in a circuit-switched telephony network, and in Chapter 4 we extended this to consider the flow of cells through an ATM buffer. In both cases, the time between 'arrivals' (whether calls or cells) was given by a negative exponential distribution: that is to say, arrivals formed a Poisson process. But although the same source model is used, different types of behaviour are being modelled. In the first case the behaviour concerns the use made of the telephony service by customers – in terms of how often the service is used, and for how long. In the second case, the focus is at the level below the call time scale, i.e. the characteristic behaviour of the service as a flow of cells or, indeed, packets. Figure 6.1 distinguishes these two different types of behaviour by considering four different time scales of activity:

- calendar: daily, weekly and seasonal variations

- connection: set-up and clear events delimit the connection duration, which is typically in the range 100 to 1000 seconds

- burst: the behaviour of a transmitting user, characterized as a cell (or packet) flow rate, over an interval during which that rate is assumed constant. For telephony, the talk-spurt on/off characteristics have durations ranging from a fraction of a second to a few seconds. In IP, similar time scales apply to packet flows.

- cell/packet: the behaviour of cell or packet generation at the lowest level, concerned with the time interval between arrivals (e.g. multiples of 2.831 µs at 155.52 Mbit/s in ATM)

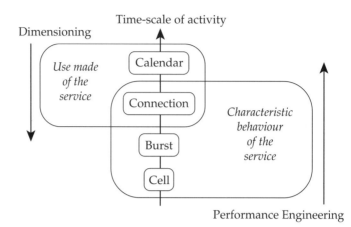

Figure 6.1. Levels of Traffic Behaviour

This analysis of traffic behaviour helps in distinguishing the primary objectives of dimensioning and performance engineering. Dimensioning focuses on the organization and provision of sufficient equipment in the network to meet the needs of services used by subscribers (i.e. at the calendar and connection levels); it does require knowledge of the service characteristics, but this is in aggregate form and not necessarily to a great level of detail. Performance engineering, however, focuses on the detail of how the network resources are able to support services (i.e. assessing the limits of performance); this requires consideration of the detail of service characteristics (primarily at the cell and burst levels), as well as information about typical service mixes – how much voice, video and data traffic is being transported on any link (which would be obtained from a study of service use).

TIMING INFORMATION IN SOURCE MODELS

A source model describes how traffic, whether cells, bursts or connections, emanates from a user. As we have already seen, the same source model can be applied to different time scales of activity, but the Poisson process is not the only one used for ATM or IP. Source models may be classified in a variety of ways: continuous time or discrete time, inter-arrival time or counting process, state-based or distribution-based, and we will consider some of these in the rest of this chapter. It is worth noting that some models are associated with a particular queue modelling method, an example being fluid flow analysis.

A distinguishing feature of source models is the way the timing information is presented. Figure 6.2 shows the three different ways in the context of an example ATM cell stream: as the number of cell slots between arrivals (the inter-arrival times are 5, 7, 3 and 5 slots in this

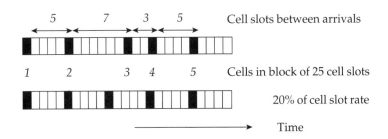

Figure 6.2. Timing Information for an Example ATM Cell Stream

example); as a count of the number of arrivals within a specified period (here, it is 5 cells in 25 cell slots); and as a cell rate, which in this case is 20% of the cell slot rate.

TIME BETWEEN ARRIVALS

Inter-arrival times can be specified by either a fixed value, or some arbitrary probability distribution of values, for the time between successive arrivals (whether cells or connections). These values may be in continuous time, taking on any real value, or in discrete time, for example an integer multiple of a discrete time period such as the transmission time of a cell, e.g. 2.831 μs.

A negative-exponential distribution of inter-arrival times is the prime example of a continuous-time process because of the 'memoryless' property. This name arises from the fact that, if the time is now t_1, the probability of there being k arrivals in the interval $t_1 \rightarrow t_2$ is independent of the interval, δt, since the last arrival (Figure 6.3). It is this property that allows the development of some of the simple formulas for queues.

The probability that the inter-arrival time is less than or equal to t is given by the equation

$$\Pr\{\text{inter-arrival time} \leqslant t\} = F(t) = 1 - e^{-\lambda \cdot t}$$

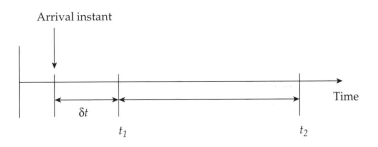

Figure 6.3. The Memoryless Property of the Negative Exponential Distribution

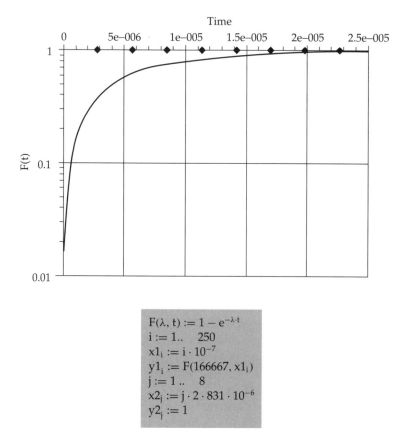

Figure 6.4. Graph of the Negative Exponential Distribution for a Load of 0.472, and the Mathcad Code to Generate (x, y) Values for Plotting the Graph

where the arrival rate is λ. This distribution, $F(t)$, is shown in Figure 6.4 for a load of 47.2% (i.e. the 1000 CBR source example from Chapter 4). The arrival rate is 166 667 cell/s which corresponds to an average inter-arrival time of 6 µs. The cell slot intervals are also shown every 2.831 µs on the time axis.

The discrete time equivalent is to have a geometrically distributed number of time slots between arrivals (Figure 6.5), where that number is counted from the end of the first cell to the end of the next cell to arrive.

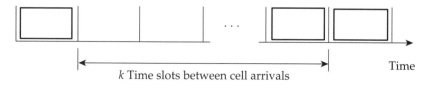

Figure 6.5. Inter-Arrival Times Specified as the Number of Time Slots between Arrivals

Obviously a cell rate of 1 cell per time slot has an inter-arrival time of 1 cell slot, i.e. no empty cell slots between arrivals. The probability that a cell time slot contains a cell is a constant, which we will call p. Hence a time slot is empty with probability $1 - p$. The probability that there are k time slots between arrivals is given by

$$\Pr\{k \text{ time slots between arrivals}\} = (1 - p)^{k-1} \cdot p$$

i.e. $k - 1$ empty time slots, followed by one full time slot. This is the geometric distribution, the discrete time equivalent of the negative exponential distribution. The geometric distribution is often introduced in text books in terms of the throwing of dice or coins, hence it is thought

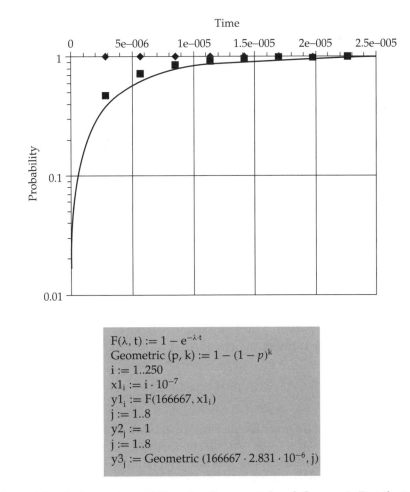

Figure 6.6. A Comparison of Negative Exponential and Geometric Distributions, and the Mathcad Code to Generate (x, y) Values for Plotting the Graph

of as having $k - 1$ 'failures' (empty time slots, to us), followed by one 'success' (a cell arrival). The mean of the distribution is the inverse of the probability of success, i.e. $1/p$. Note that the geometric distribution also has a 'memoryless' property in that the value of p for time slot n remains constant however many arrivals there have been in the previous $n - 1$ slots.

Figure 6.6 compares the geometric and negative exponential distributions for a load of 47.2% (i.e. for the geometric distribution, $p = 0.472$, with a time base of 2.831 µs; and for the negative exponential distribution, $\lambda = 166\,667$ cell/s, as before). These are cumulative distributions (like Figure 6.4), and they show the probability that the inter-arrival time is less than or equal to a certain value on the time axis. This time axis is sub-divided into cell slots for ease of comparison. The cumulative geometric distribution begins at time slot $k = 1$ and adds $\Pr\{k$ time slots between arrivals$\}$ for each subsequent value of k.

$$\Pr\{\leqslant k \text{ time slots between arrivals}\} = 1 - (1 - p)^k$$

COUNTING ARRIVALS

An alternative way of presenting timing information about an arrival process is by counting the number of arrivals in a defined time interval. There is an equivalence here with the inter-arrival time approach in continuous time: negative exponential distributed inter-arrival times form a Poisson process:

$$\Pr\{k \text{ arrivals in time } T\} = \frac{(\lambda \cdot T)^k}{k!} \cdot e^{-\lambda \cdot T}$$

where λ is the arrival rate.

In discrete time, geometric inter-arrival times form a Bernoulli process, where the probability of one arrival in a time slot is p and the probability of no arrival in a time slot is $1 - p$. If we consider more than one time slot, then the number of arrivals in N slots is binomially distributed:

$$\Pr\{k \text{ arrivals in } N \text{ time slots}\} = \frac{N!}{(N - k)! \cdot k!} \cdot (1 - p)^{N-k} \cdot p^k$$

and p is the average number of arrivals per time slot.

How are these distributions used to model ATM or IP systems? Consider the example of an ATM source that is generating cell arrivals as a Poisson process; the cells are then buffered, and transmitted in the usual way for ATM – as a cell stream in synchronized slots (see Figure 6.7). The Poisson process represents cells arriving from the source

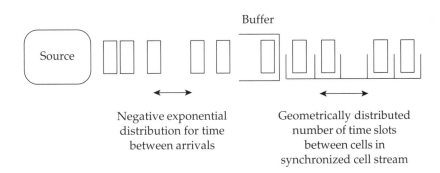

Figure 6.7. The Bernoulli Output Process as an *Approximation* to a Poisson Arrival Stream

to the buffer, at a cell arrival rate of λ cells per time slot. At the buffer output, a cell occupies time slot i with probability p as we previously defined for the Bernoulli process. Now if λ is the cell arrival rate and p is the output cell rate (both in terms of number of cells per time slot), and if we are not losing any cells in our (infinite) buffer, we must have that $\lambda = p$.

Note that the output process of an ATM buffer of infinite length, fed by a Poisson source is *not* actually a Bernoulli process. The reason is that the queue introduces dependence from slot to slot. If there are cells in the buffer, then the probability that no cell is served at the next cell slot is 0, whereas for the Bernoulli process it is $1 - p$. So, although the output cell stream is not a memoryless process, the Bernoulli process is still a useful approximate model, variations of which are frequently encountered in teletraffic engineering for ATM and for IP.

The limitation of the negative exponential and geometric inter-arrival processes is that they do not incorporate all of the important characteristics of typical traffic, as will become apparent later.

Certain forms of switch analysis assume 'batch-arrival' processes: here, instead of a single arrival with probability p, we get a group (the batch), and the number in the group can have any distribution. This form of arrival process can also be considered in this category of counting arrivals. For example, at a buffer in an ATM switch, a batch of arrivals up to some maximum, M, arrive from different parts of the switch during a time slot. This can be thought of as counting the same number of arrivals as cells in the batch during that time slot. The Bernoulli process with batch arrivals is characterized by having an independent and identically distributed number of arrivals per discrete time period. This is defined in two parts: the presence of a batch

$$\Pr\{\text{there is a batch of arrivals in a time slot}\} = p$$

or the absence of a batch

$$\text{Pr}\{\text{there is no batch of arrivals in a time slot}\} = 1 - p$$

and the distribution of the number of cells in a batch:

$b(k) = \text{Pr}\{\text{there are } k \text{ cells in a batch given that there is a batch in the}$

$\times$ time slot$\}$

Note that k is greater than 0. This description of the arrival process can be rearranged to give the overall distribution of the number of arrivals per slot, $a(k)$, as follows:

$$a(0) = 1 - p$$
$$a(1) = p \cdot b(1)$$
$$a(2) = p \cdot b(2)$$

$$\vdots$$

$$a(k) = p \cdot b(k)$$

$$\vdots$$

$$a(M) = p \cdot b(M)$$

This form of input is used in the switching analysis described in Chapter 7 and the basic packet queueing analysis described in Chapter 14. It is a general form which can be used for both Poisson and binomial input distributions, as well as arbitrary distributions. Indeed, in Chapter 17 we use a batch arrival process to model long-range dependent traffic, with Pareto-distributed batch sizes.

In the case of a Poisson input distribution, the time duration T is one time slot, and if λ is the arrival rate in cells per time slot, then

$$a(k) = \frac{\lambda^k}{k!} \cdot e^{-\lambda}$$

For the binomial distribution, we now want the probability that there are k arrivals from M inputs where each input has a probability, p, of producing a cell arrival in any time slot. Thus

$$a(k) = \frac{M!}{(M - k)! \cdot k!} \cdot (1 - p)^{M-k} \cdot p^k$$

and the total arrival rate is $M \cdot p$ cells per time slot. Figure 6.8 shows what happens when the total arrival rate is fixed at 0.95 cells per time

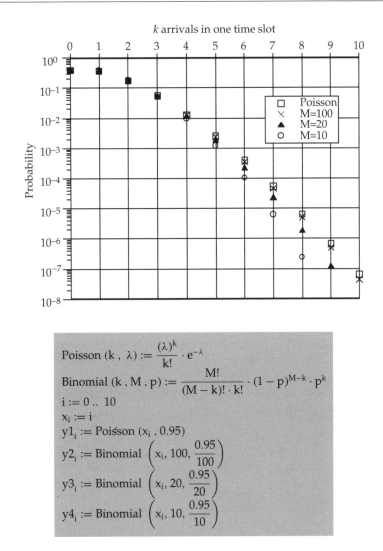

Figure 6.8. A Comparison of Binomial and Poisson Distributions, and the Mathcad Code to Generate (x, y) Values for Plotting the Graph

slot and the numbers of inputs are 10, 20 and 100 (and so p is 0.095, 0.0475 and 0.0095 respectively). The binomial distribution tends towards the Poisson distribution, and in fact in the limit as $N \to \infty$ and $p \to 0$ the distributions are the same.

RATES OF FLOW

The simplest form of source using a rate description is the periodic arrival stream. We have already met an example of this in 64 kbit/s CBR telephony, which has a cell rate of 167 cell/s in ATM. The next step is to consider an ON–OFF source, where the process switches between a silent state, producing no cells, and a state which produces a particular fixed rate of cells. Sources with durations (in the ON and OFF states) distributed as negative exponentials have been most frequently studied, and have been applied to data traffic, to packet-speech traffic, and as a general model for bursty traffic in an ATM multiplexor.

Figure 6.9 shows a typical teletraffic model for an ON–OFF source. During the time in which the source is on (called the 'sojourn time in the active state'), the source generates cells at a rate of R. After each cell, another cell is generated with probability a, or the source changes to the silent state with probability $1 - a$. Similarly, in the silent state, the source generates another empty time slot with probability s, or moves to the active state with probability $1 - s$. This type of source generates cells in patterns like that shown in Figure 6.10; for this pattern, R is equal to half of the cell slot rate. Note that there are empty slots during the active state; these occur if the cell arrival rate, R, is less than the cell slot rate.

We can view the ON–OFF source in a different way. Instead of showing the cell generation process and empty time slot process explicitly as Bernoulli processes, we can simply describe the active state as having a geometrically distributed number of cell arrivals, and the silent state as having a geometrically distributed number of cell slots. The mean number of cells in an active state, E[on], is equal to the inverse of the probability of exiting the active state, i.e. $1/(1 - a)$ cells. The mean number of empty

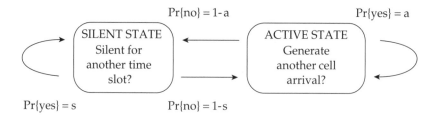

Figure 6.9.   An ON–OFF Source Model

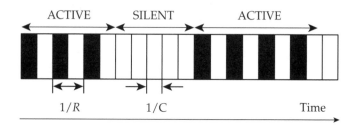

Figure 6.10.   Cell Pattern for an ON–OFF Source Model

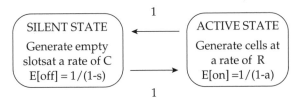

Figure 6.11. An Alternative Representation of the ON–OFF Source Model

cell slots in a silent state, E[off], is equal to $1/(1-s)$ cell slots. At the end of a sojourn period in a state, the process switches to the other state with probability 1. Figure 6.11 shows this alternative representation of the ON–OFF source model.

It is important to note that the geometric distributions for the active and silent states have different time bases. For the active state the unit of time is $1/R$, i.e. the cell inter-arrival time. Thus the mean duration in the active state is

$$T_{on} = \frac{1}{R} \cdot E[on]$$

For the silent state the unit of time is $1/C$, where C is the cell slot rate; thus the mean duration in the silent state is

$$T_{off} = \frac{1}{C} \cdot E[off]$$

The alternative representation of Figure 6.11 can then be generalized by allowing arbitrary distributions for the number of cells generated in an active period, and also for the number of empty slots generated in a silent period.

Before leaving the ON–OFF source, let's apply it to a practical example: silence-suppressed telephony (no cells are transmitted during periods in which the speaker is silent). Typical figures (found by measurement) for the mean ON and OFF periods are 0.96 second and 1.69 seconds respectively. Cells are generated from a 64 kbit/s telephony source at a rate of $R = 167$ cell/s and the cell slot rate of a 155.52 Mbit/s link is $C = 353\,208$ cell/s. Thus the mean number of cells produced in an active state is

$$E[on] = R \times 0.96 = 160 \text{ cells}$$

and the mean number of empty slots in a silent state is

$$E[off] = C \times 1.69 = 596\,921 \text{ cell slots}$$

This gives the model shown in Figure 6.12.

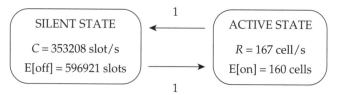

Figure 6.12.   ON–OFF Source Model for Silence-Suppressed Telephony

We can also calculate values of parameters a and s for the model in Figure 6.9. We know that

$$E[on] = \frac{1}{1-a} = 160$$

so

$$a = 1 - \frac{1}{160} = 0.993\,75$$

and

$$E[off] = \frac{1}{1-s} = 596\,921$$

so

$$s = 1 - \frac{1}{596\,921} = 0.999\,998\,324\,7$$

The ON–OFF source is just a particular example of a state-based model in which the arrival rate in a state is fixed, there are just two states, and the period of time spent in a state (the sojourn time) is negative exponentially, geometrically, or arbitrarily distributed. We can generalize this to incorporate N states, with fixed rates in each state. These multi-state models (called 'modulated deterministic processes') are useful for modelling a number of ON–OFF sources multiplexed together, or a single, more complex, traffic source such as video.

If we allow the sojourn times to have arbitrary distributions, the resulting process is called a Generally Modulated Deterministic Process (GMDP). If the state durations are exponentially distributed then the process is called a Markov Modulated Deterministic Process (MMDP). In this case, each state produces a geometrically distributed number of cells during any sojourn period. This is because, having generated arrival i, it generates arrival $i + 1$ with a probability given by the probability that the sojourn time does not end before the time of the next arrival. This probability is a constant if sojourn periods are exponentially distributed because of the 'memoryless' property of the negative exponential distribution.

We do not need to restrict the model to having a constant arrival rate in each state: if the arrival process per state is a Poisson process, and the

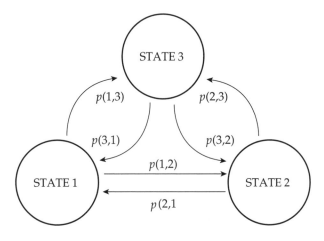

Figure 6.13. The Three-State GMDP

mean of the Poisson distribution is determined by the state the model is in, then we have an MMPP, which is useful for representing an aggregate cell arrival process.

For all these state processes, at the end of a sojourn in state i, a transition is made to another state j; this transition is governed by an $N \times N$ matrix of transition probabilities, $p(i, j)$ $i \neq j$. Figure 6.13 illustrates a multi-state model, with three states, and with the transition probabilities from state i to state j shown as $p(i, j)$.

For a comprehensive review of traffic models, the reader is referred to [6.1].

PART II

ATM Queueing and Traffic Control

7 Basic Cell Switching

up against the buffers

THE QUEUEING BEHAVIOUR OF ATM CELLS IN OUTPUT BUFFERS

In Chapter 3, we saw how teletraffic engineering results have been used to dimension circuit-switched telecommunications networks. ATM is a connection-orientated telecommunications network, and we can (correctly) anticipate being able to use these methods to investigate the connection-level behaviour of ATM traffic. However, the major difference between circuit-switched networks and ATM is that ATM connections consist of a cell stream, where the time between these cells will usually be variable (at whichever point in the network that you measure them). We now need to consider what may happen to such a cell stream as it travels through an ATM switch (it will, in general, pass through many such switches as it crosses the network).

The purpose of an ATM switch is to route arriving cells to the appropriate output. A variety of techniques have been proposed and developed to do switching [7.1], but the most common uses output buffering. We will therefore concentrate our analysis on the behaviour of the output buffers in ATM switches. There are three different types of behaviour in which we are interested: the state probabilities, by which we mean the proportion of time that a queue is in a particular state (being in state k means the queue contains k cells) over a very long period of time (i.e. the *steady-state* probabilities); the cell loss probability, by which we mean the proportion of cells lost over a very long period of time; and the cell waiting-time probabilities, by which we mean the probabilities associated with a cell being delayed k time slots.

To analyse these different types of behaviour, we need to be aware of the timing of events in the output buffer. In ATM, the cell service is of fixed duration, equal to a single time slot, and synchronized so that a cell

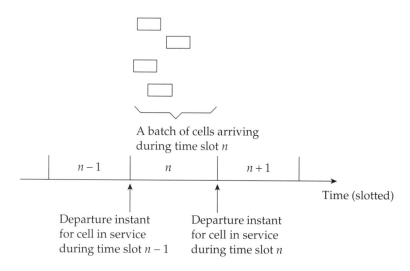

A batch of cells arriving
during time slot n

| $n-1$ | n | $n+1$ |

Time (slotted)

Departure instant   Departure instant
for cell in service   for cell in service
during time slot $n-1$ during time slot n

Figure 7.1. Timing of Events in the Buffer: the Arrivals-First Buffer Management Strategy

enters service at the beginning of a time slot. The cell departs at the end of a time slot, and this is synchronized with the start of service of the next cell (or empty time slot, if there is nothing waiting in the buffer). Cells arrive during time slots, as shown in Figure 7.1. The exact instants of arrival are unimportant, but we will assume that any arrivals in a time slot occur before the departure instant for the cell in service during the time slot. This is called an 'arrivals-first' buffer management strategy. We will also assume that if a cell arrives during time slot n, the earliest it can be transmitted (served) is during time slot $n+1$.

For our analysis, we will use a Bernoulli process with batch arrivals, characterized by an independent and identically distributed batch of k arrivals ($k = 0, 1, 2, \ldots$) in each cell slot:

$$a(k) = \Pr\{k \text{ arrivals in a cell slot}\}$$

It is particularly important to note that the state probabilities refer to the state of the queue at moments in time that are usually called the 'end of time-slot instants'. These instants are after the arrivals (if there are any) and after the departure (if there is one); indeed they are usually defined to be at a time Δt after the end of the slot, where $\Delta t \to 0$.

BALANCE EQUATIONS FOR BUFFERING

The effect of random arrivals on the queue is shown in Figure 7.2. For the buffer to contain i cells *at the end of any time slot* it could have contained any one of $0, 1, \ldots, i+1$ *at the end of the previous slot*. State i can be reached

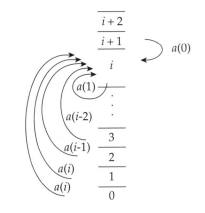

Figure 7.2. How to Reach State i at the End of a Time Slot from States at the End of the Previous Slot

from any of the states 0 up to i by a precise number of arrivals, i down to 1 (with probability $a(i)\ldots a(1)$) as expressed in the figure (note that not all the transitions are shown). To move from $i+1$ to i requires that there are no arrivals, the probability of which is expressed as $a(0)$; this then reflects the completion of service of a cell during the current time slot.

We define the state probability, i.e. the probability of being in state k, as

$$s(k) = \Pr\{\text{there are } k \text{ cells in the queueing system at the end of any}$$

$$\times \text{ time slot}\}$$

and again (as in Chapter 4) we begin by making the simplifying assumption that the queue has infinite capacity. This means we can find the 'system empty' probability, $s(0)$ from simple traffic theory. We know from Chapter 3 that

$$L = A - C$$

where L is the lost traffic, A is the offered traffic and C is the carried traffic. But if the queue is infinite, then there is no loss ($L = 0$), so

$$A = C$$

This time, though, we are dealing with a stream of cells, not calls. Thus our offered traffic is numerically equal to λ, the mean arrival rate of cells in cell/s (because the cell service time, s, is one time slot), and the carried traffic is the mean number of cells served per second, i.e. it is the utilization divided by the service time per cell, so

$$\lambda = \frac{\rho}{s}$$

If we now consider the service time of a cell to be one time slot, for simplicity, then the average number of arrivals per time slot is denoted $E[a]$ (which is the mean of the arrival distribution $a(k)$), and the average number of cells carried per time slot is the utilization. Thus

$$E[a] = \rho$$

But the utilization is just the steady-state probability that the system is not empty, so

$$E[a] = \rho = 1 - s(0)$$

and therefore

$$s(0) = 1 - E[a]$$

So from just the arrival rate (without any knowledge of the arrival distribution $a(k)$) we are able to determine the probability that the system is empty at the end of any time slot. It is worth noting that, if the applied cell arrival rate is greater than the cell service rate (one cell per time slot), then

$$s(0) < 0$$

which is a very silly answer! Obviously then we need to ensure that cells are not arriving faster (on average) than the system is able to transmit them. If $E[a] \geqslant 1$ cell per time slot, then it is said that the queueing system is *unstable*, and the number of cells in the buffer will simply grow in an unbounded fashion.

CALCULATING THE STATE PROBABILITY DISTRIBUTION

We can build on this value, $s(0)$, by going back to the idea of adding all the ways in which it is possible to end up in any particular state. Starting with state 0 (the system is empty), this can be reached from a system state of either 1 or 0, as shown in Figure 7.3. This is saying that the system can be in state 0 at the end of slot $n - 1$, with no arrivals in slot n, or it can be in state 1 at the end of slot $n - 1$, with no arrivals in slot n, and at the end of slot n, the system will be in state 0.

We can write an equation to express this relationship:

$$s(0) = s(0) \cdot a(0) + s(1) \cdot a(0)$$

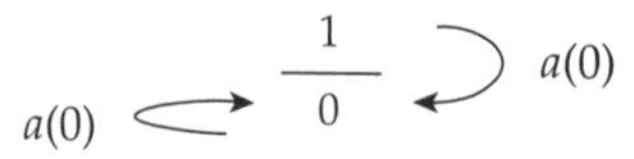

Figure 7.3.   How to Reach State 0 at the End of a Time Slot

You may ask how it can be that $s(k)$ applies as the state probabilities for the end of time slot $n - 1$ and time slot n. Well, the answer lies in the fact that these are steady-state (sometimes called 'long-run') probabilities, and, on the assumption that the buffer has been active for a very long period, the probability distribution for the queue at the end of time slot $n - 1$ is the same as the probability distribution for the end of time slot n. Our equation can be rearranged to give a formula for $s(1)$:

$$s(1) = s(0) \cdot \frac{1 - a(0)}{a(0)}$$

In a similar way, we can find a formula for $s(2)$ by writing a balance equation for $s(1)$:

$$s(1) = s(0) \cdot a(1) + s(1) \cdot a(1) + s(2) \cdot a(0)$$

Again, this is expressing the probability of having 1 in the queueing system at the end of slot n, in terms of having 0, 1 or 2 in the system at the end of slot $n - 1$, along with the appropriate number of arrivals (Figure 7.4). Remember, though, that any arrivals during the current time slot cannot be served during this slot.

Rearranging the equation gives:

$$s(2) = \frac{s(1) - s(0) \cdot a(1) - s(1) \cdot a(1)}{a(0)}$$

We can continue with this process to find a similar expression for the general state, k.

$$s(k - 1) = s(0) \cdot a(k - 1) + s(1) \cdot a(k - 1) + s(2) \cdot a(k - 2) + \cdots + s(k - 1)$$
$$\cdot a(1) + s(k) \cdot a(0)$$

which, when rearranged, gives:

$$s(k) = \frac{s(k - 1) - s(0) \cdot a(k - 1) - \sum_{i=1}^{k-1} s(i) \cdot a(k - i)}{a(0)}$$

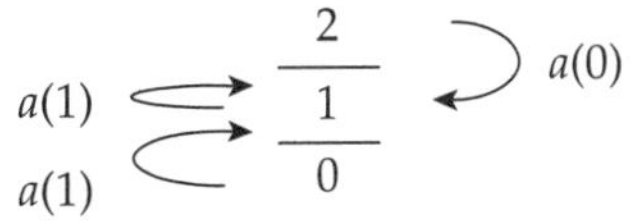

Figure 7.4. How to Reach State 1 at the End of a Time Slot

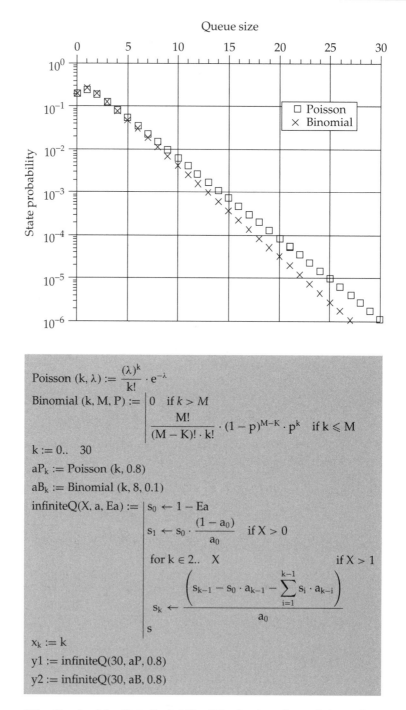

Figure 7.5. Graph of the State Probability Distributions for an Infinite Queue with Binomial and Poisson Input, and the Mathcad Code to Generate (x, y) Values for Plotting the Graph

Because we have used the simplifying assumption that the queue length is infinite, we can, theoretically, make k as large as we like. In practice, how large we can make it will depend upon the value of $s(k)$ that results from this calculation, and the program used to implement this algorithm (depending on the relative precision of the real-number representation being used).

Now what about results? What does this state distribution look like? Well, in part this will depend on the actual input distribution, the values of $a(k)$, so we can start by obtaining results for the two input distributions discussed in Chapter 6: the binomial and the Poisson. Specifically, let us

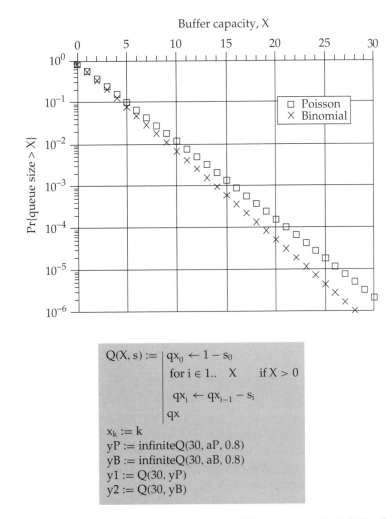

$$Q(X, s) := \begin{vmatrix} qx_0 \leftarrow 1 - s_0 \\ \text{for } i \in 1.. \ X \qquad \text{if } X > 0 \\ \\ qx_i \leftarrow qx_{i-1} - s_i \\ qx \end{vmatrix}$$

$x_k := k$
$yP := \text{infiniteQ}(30, aP, 0.8)$
$yB := \text{infiniteQ}(30, aB, 0.8)$
$y1 := Q(30, yP)$
$y2 := Q(30, yB)$

Figure 7.6. Graph of the Approximation to the Cell Loss by the Probability that the Queue State Exceeds X, and the Mathcad Code to Generate (x, y) Values for Plotting the Graph

assume an output-buffered switch, and plot the state probabilities for an infinite queue at one of the output buffers; the arrival rate per input is 0.1 (i.e. the probability that an input port contains a cell destined for the output buffer in question is 0.1 for any time slot) and $M = 8$ input and output ports. Thus we have a binomial distribution with parameters $M = 8, p = 0.1$, compared to a Poisson distribution with mean arrival rate of $M \cdot p = 0.8$ cells per time slot. Both are shown in Figure 7.5.

What then of cell loss? Well, with an infinite queue we will not actually have any; in the next section we will deal exactly with the cell loss probability (CLP) from a finite queue of capacity X. Before we do so, it is worth considering approximations for the CLP found from the infinite buffer case. As with Chapter 4, we can use the probability that there are more than X cells in the infinite buffer as an approximation for the CLP. In Figure 7.6 we plot this value, for both the binomial and Poisson cases considered previously, over a range of buffer length values.

EXACT ANALYSIS FOR FINITE OUTPUT BUFFERS

Having considered infinite buffers, we now want to quantify exactly the effect of a finite buffer, such as we would actually find acting as the output buffer in a switch. We want to know how the CLP at this queue varies with the buffer capacity, X, and to do this we need to use the balance equation technique. However, this time we cannot find $s(0)$ directly, by equating carried traffic and offered traffic, because there will be some lost traffic, and it is this that we need to find!

So initially we use the same approach as for the infinite queue, temporarily ignoring the fact that we do not know $s(0)$:

$$s(1) = s(0) \cdot \frac{1 - a(0)}{a(0)}$$

$$s(k) = \frac{s(k-1) - s(0) \cdot a(k-1) - \sum_{i=1}^{k-1} s(i) \cdot a(k-i)}{a(0)}$$

For the system to become full with the 'arrivals-first' buffer management strategy, there is actually only one way in which this can happen *at the end of time-slot instants*: to be full at the end of time slot i, the buffer must begin slot i empty, and have X or more cells arrive in the slot. If the system is non-empty at the start, then just before the end of the slot (given enough arrivals) the system will be full, but when the cell departure occurs at the slot end, there will be $X - 1$ cells left, and *not* X. So for the full state, we have:

$$s(X) = s(0) \cdot A(X)$$

where

$$A(k) = 1 - a(0) - a(1) - \cdots - a(k-1)$$

So $A(k)$ is the probability that at least k cells arrive in a slot. Now we face the problem that, without the *value* for $s(0)$, we cannot evaluate $s(k)$ for $k > 0$. What we do is to define a new variable, $u(k)$, as follows:

$$u(k) = \frac{s(k)}{s(0)}$$

so

$$u(0) = 1$$

Then

$$u(1) = \frac{1 - a(0)}{a(0)}$$

$$u(k) = \frac{u(k-1) - a(k-1) - \displaystyle\sum_{i=1}^{k-1} u(i) \cdot a(k-i)}{a(0)}$$

$$u(X) = A(X)$$

and all the values of $u(k)$, $0 \leqslant k \leqslant X$, can be evaluated! Then using the fact that all the state probabilities must sum to 1, i.e.

$$\sum_{i=0}^{X} s(i) = 1$$

we have

$$\sum_{i=0}^{X} \frac{s(i)}{s(0)} = \frac{1}{s(0)} = \sum_{i=0}^{X} u(i)$$

so

$$s(0) = \frac{1}{\displaystyle\sum_{i=0}^{X} u(i)}$$

The other values of $s(k)$, for $k > 0$, can then be found from the definition of $u(k)$:

$$s(k) = s(0) \cdot u(k)$$

Now we can apply the basic traffic theory again, using the relationship between offered, carried and lost traffic at the *cell* level, i.e.

$$L = A - C$$

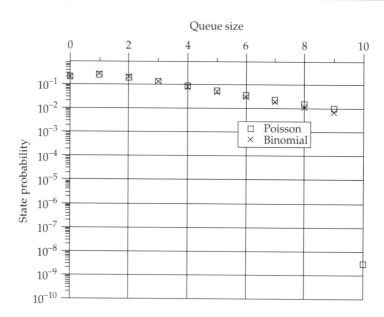

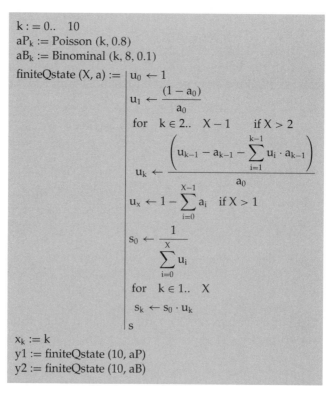

Figure 7.7. Graph of the State Probability Distribution for a Finite Queue of 10 Cells and a Load of 80%, and the Mathcad Code to Generate (x, y) Values for Plotting the Graph

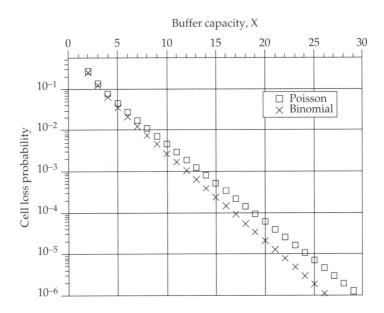

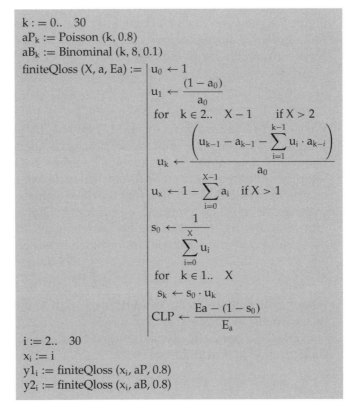

Figure 7.8. Graph of the Exact Cell Loss Probability against System Capacity X for a Load of 80%

As before, we consider the service time of a cell to be one time slot, for simplicity; then the average number of arrivals per time slot is $E[a]$ and the average number of cells carried per time slot is the utilization. Thus

$$L = E[a] - \rho = E[a] - (1 - s(0))$$

and the cell loss probability is just the ratio of lost traffic to offered traffic:

$$CLP = \frac{E[a] - (1 - s(0))}{E[a]}$$

Figure 7.7 shows the state probability distribution for an output buffer of capacity 10 cells (which includes the server) being fed from our 8 Bernoulli sources each having $p = 0.1$ as before. The total load is 80%. Notice that the probability of the buffer being full is very low in the Poisson case, and zero in the binomial case. This is because the arrivals-first strategy needs 10 cells to arrive at an empty queue in order for the queue to fill up; the maximum batch size with 8 Bernoulli sources is 8 cells.

Now we can generate the exact cell loss probabilities for finite buffers. Figure 7.8 plots the exact CLP value for binomial and Poisson input to a finite queue of system capacity X, where X varies from 2 up to 30 cells. Now compare this with Figure 7.6.

DELAYS

We looked at waiting times in M/M/1 and M/D/1 queueing systems in Chapter 4. Waiting time plus service time gives the system time, which is the overall delay through the queueing system. So, how do we work out the probabilities associated with particular delays in the output buffers of an ATM switch? Notice first that the delay experienced by a cell, which we will call cell **C**, in a buffer has two components: the delay due to the 'unfinished work' (cells) in the buffer when cell **C** arrives, U_d; and the delay caused by the other cells in the batch in which **C** arrives, B_d.

$$T_d = U_d + B_d$$

where T_d is the total delay from the arrival of **C** until the completion of its transmission (the total system time).

In effect we have already determined U_d; these values are given by the state probabilities as follows:

$$\Pr\{U_d = 1\} = U_d(1) = s(0) + s(1)$$

Remember that we assumed that each cell will be delayed by at least 1 time slot, the slot in which it is transmitted. For all $k > 1$ we have the relationship:

$$\Pr\{U_d = k\} = U_d(k) = s(k)$$

The formula for $B_d(k) = \Pr\{B_d = k\}$ accounts for the position of **C** within the batch as well:

$$B_d(k) = \frac{1 - \displaystyle\sum_{i=0}^{k} a(i)}{E[a]}$$

Note that this equation is covered in more depth in Chapter 13.

Now the total delay, $T_d(k)$, consists of all the following possibilities:

$$T_d(k) = \Pr\{U_d = 1 \text{ and } B_d = k - 1\} + \Pr\{U_d = 2 \text{ and } B_d = k - 2\} + \cdots$$

and we account for them all by convolving the two components of delay, using the following formula:

$$T_d(k) = \sum_{j=1}^{k} U_d(j) \cdot B_d(k - j)$$

We plot the cell delay probabilities for the example we have been considering (binomial and Poisson input processes, $p = 0.1$ and $M = 8$, $\rho = 0.8$) in Figure 7.9.

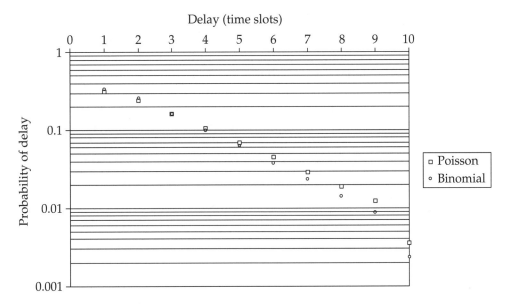

Figure 7.9. Cell Delay Probabilities for a Finite Buffer of Size 10 Cells with a Load of 80%

End-to-end delay

To find the cell delay variation through a number of switches, we convolve the cell delay distribution for a single buffer with itself. Let

$$T_{d,n}(k) = \text{Pr}\{\text{total delay through } n \text{ buffers} = k\}$$

Then, for two switches the delay distribution is given by

$$T_{d,2}(k) = \sum_{j=1}^{k} T_{d,1}(j) \cdot T_{d,1}(k-j)$$

There is one very important assumption we are making: that the arrivals to each buffer are independent of each other. This is definitely *not* the case if all the traffic through the first buffer goes through the second one. In practice, it is likely that only a small proportion will do so; the bulk of the traffic will be routed elsewhere. This situation is shown in Figure 7.10.

We can extend our calculation for 2 switches by applying it recursively to find the delay through n buffers:

$$T_{d,n}(k) = \sum_{j=1}^{k} T_{d,n-1}(j) \cdot T_{d,1}(k-j)$$

Figure 7.11 shows the end-to-end delay distributions for 1, 2, 3, 5, 7 and 9 buffers, where the buffers have identical but independent binomial arrival distributions, each buffer is finite with a size of 10 cells, and the load offered to each buffer is 80%. Lines are shown as well as markers on the graph to help identify each distribution; obviously, the delay can only take integer values. As we found in Chapter 4, the delay distribution 'flattens' as the number of buffers increases. Note that this is a *delay* distribution, which includes one time slot for the server in each buffer; in Figure 4.8, it is the end-to-end *waiting time*

Figure 7.10. Independence Assumption for End-to-End Delay Distribution: 'Through' Traffic is a Small Proportion of Total Traffic Arriving at Each Buffer

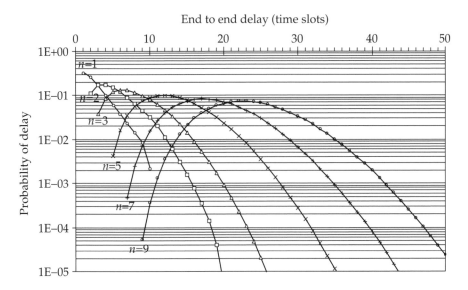

Figure 7.11. End-to-End Delay Distributions for 1, 2, 3, 5, 7 and 9 Buffers, with a Load of 80%

distribution which is shown. So, for example, in the distribution for end-to-end delay through 9 buffers, the smallest delay is 9 time slots (and the largest delay is 90 time slots, although this is not shown in Figure 7.11).

8 Cell-Scale Queueing

dealing with the jitters

CELL-SCALE QUEUEING

In Chapter 4 we considered a situation in which a large collection of CBR voice sources all send their cells to a single buffer. We stated that it was reasonably accurate under certain circumstances (when the number of sources is large enough) to model the total cell-arrival process from all the voice sources as a Poisson process.

Now a Poisson process is a single statistical model from which the detailed information about the behaviour of the individual sources has been lost, quite deliberately, in order to achieve simplicity. The process features a random number (a batch) of arrivals per slot (see Figure 8.1) where this batch can vary as $0, 1, 2, \ldots, \infty$.

So we could say that in, for example, slot $n + 4$, the process has overloaded the queueing system because two cells have arrived – one more than the buffer can transmit. Again, in slot $n + 5$ the buffer has been overloaded by three cells in the slot. So the process provides short periods during which its instantaneous arrival rate is greater than the cell service rate; indeed, if this did not happen, there would be no need for a buffer.

But what does this mean for our N CBR sources? Each source is at a *constant* rate of 167 cell/s, so the cell rate will never individually exceed the service rate of the buffer; and provided $N \times 167 < 353\,208$ cell/s, the total cell rate will not do so either. The maximum number of sources is $353\,208/167 = 2115$ or, put another way, each source produces one cell every 2115 time slots. However, the sources are not necessarily arranged such that a cell from each one arrives in its own time slot; indeed, although the probability is not high, all the sources could be (accidentally) synchronized such that all the cells arrive in the same slot. In fact, for our example of multiplexing 2115 CBR sources, it is possible

Number of arrivals in a slot

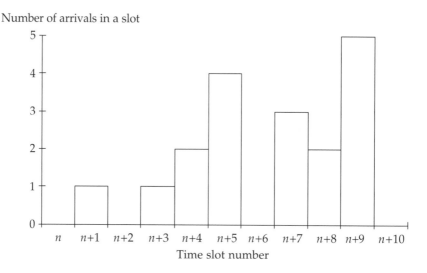

Figure 8.1.   A Random Number of Arrivals per Time Slot

for any number of cells varying from 0 up to 2115 to arrive in the same slot. The queueing behaviour which arises from this is called 'cell-scale queueing'.

MULTIPLEXING CONSTANT-BIT-RATE TRAFFIC

Let us now take a closer look at what happens when we have constant-bit-rate traffic multiplexed together. Figure 8.2 shows, for a simple situation, how repeating patterns develop in the arrival process – patterns which depend on the relative phases of the sources.

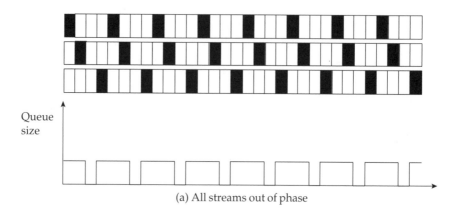

(a) All streams out of phase

Figure 8.2.   Repeating Patterns in the Size of the Queue when Constant-Bit-Rate Traffic Is Multiplexed

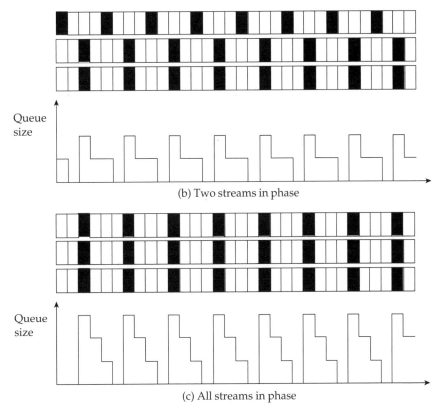

(b) Two streams in phase

(c) All streams in phase

Figure 8.2.   (*continued*)

It is clear from this picture that there are going to be circumstances where a simple 'classical' queueing system like the M/D/1 will not adequately model superposed CBR traffic; in particular, the arrival process is not well modelled by a Poisson process when the number of sources is small. At this point we need a fresh start with a new approach to the analysis.

ANALYSIS OF AN INFINITE QUEUE WITH MULTIPLEXED CBR INPUT: THE *N·D/D/1*

The N·D/D/1 queue is a basic model for CBR traffic where the input process comprises N independent periodic sources, each source with the same period D. If we take our collection of 1000 CBR sources, then $N = 1000$, and $D = 2115$ time slots. The queueing analysis caters for all possible repeating patterns and their effect on the queue size. The buffer capacity is assumed to be infinite, and the cell loss probability is approximated by the probability that the queue exceeds a certain size x, i.e. $Q(x)$. Details of the derivation can be found in [8.1].

$$\text{CLP} \approx Q(x) = \sum_{n=x+1}^{N} \left\{ \frac{N!}{n! \cdot (N-n)!} \cdot \left(\frac{n-x}{D}\right)^{n} \cdot \left[1 - \left(\frac{n-x}{D}\right)\right]^{N-n} \right.$$
$$\left. \cdot \frac{D-N+x}{D-n+x} \right\}$$

Let's put some numbers in, and see how the cell loss varies with different parameters and their values. The distribution of $Q(x)$ for a fixed load of

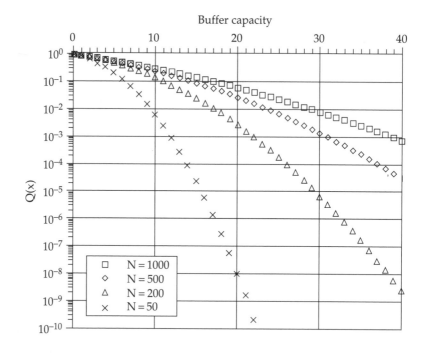

Figure 8.3. Results for the N·D/D/1 Queue with a Load of 95%, and the Mathcad Code to Generate (x, y) Values for Plotting the Graph

$\rho = N/D = 0.95$ with numbers of sources ranging from 50 up to 1000 is given in Figure 8.3. Note how the number of inputs (sources) has such a significant impact on the results. Remember that the traffic is periodic, and the utilization is less than 1, so the maximum number of arrivals in any one period of the constant-bit-rate sources (as well as in any one time slot) is limited to one from each source, i.e. N. The value of N limits the maximum size of the queue – if we provide N waiting spaces there would be no loss at all.

The N·D/D/1 result can be simplified when the applied traffic is close to the service rate; this is called a 'heavy traffic theorem'. But let's first look at a useful heavy traffic result for a queueing system we already know – the M/D/1.

HEAVY-TRAFFIC APPROXIMATION FOR THE M/D/1 QUEUE

An approximate analysis of the M/D/1 system produces the following equation:

$$Q(x) = e^{-2 \cdot x \cdot \left(\frac{1-\rho}{\rho}\right)}$$

Details of the derivation can be found in [8.2]. The result amounts to approximating the queue length by an exponential distribution: $Q(x)$ is the probability that the queue size exceeds x, and ρ is the utilization. At first sight, this does not seem to be reasonable; the number in the queue is always an integer, whereas the exponential distribution applies to a continuous variable x; and although x can vary from zero up to infinity, we are using it to represent a finite buffer size. However, it does work: $Q(x)$ is a good approximation for the cell loss probability for a finite buffer of size x. In later chapters we will develop equations for $Q(x)$ for discrete distributions.

For this equation to be accurate, the utilization must be high. Figure 8.4 shows how it compares with our exact analysis from Chapter 7, with Poisson input traffic at different values of load. The approximate results are shown as lines through the origin. It is apparent that although the cell loss approximation safely overestimates at high utilization, it can significantly underestimate when the utilization is low. But in spite of this weakness, the major contribution that this analysis makes is to show that there is a log–linear relationship between cell loss probability and buffer capacity.

Why is this heavy-traffic approximation so useful? We can rearrange the equation to specify any one variable in terms of the other two. Recalling the conceptual framework of the traffic–capacity–performance model from Chapter 3, we can see that the traffic is represented by ρ (the utilization), the capacity is x (the buffer size), and the performance

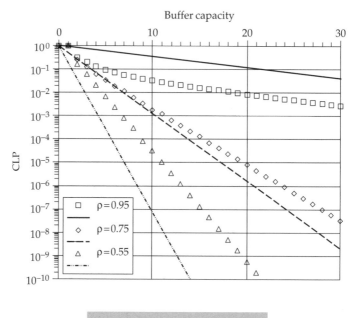

Figure 8.4. Comparing the Heavy-Traffic Results for the M/D/1 with Exact Analysis of the M/D/1/K, and the Mathcad Code to Generate (x, y) Values for Plotting the Graph

is $Q(x)$ (the approximation to the cell loss probability). Taking natural logarithms of both sides of the equation gives

$$\ln(Q(x)) = -2x\frac{(1-\rho)}{\rho}$$

This can be rearranged to give

$$x = -\frac{1}{2}\ln(Q(x))\left(\frac{\rho}{1-\rho}\right)$$

and

$$\rho = \frac{2x}{2x - \ln(Q(x))}$$

We will not investigate how to use these equations just yet. The first relates to buffer dimensioning, and the second to admission control, and both these topics are dealt with in later chapters.

HEAVY-TRAFFIC APPROXIMATION FOR THE *N·D/D/1* QUEUE

Although the exact solution for the N·D/D/1 queue is relatively straightforward, the following heavy-traffic approximation for the N·D/D/1

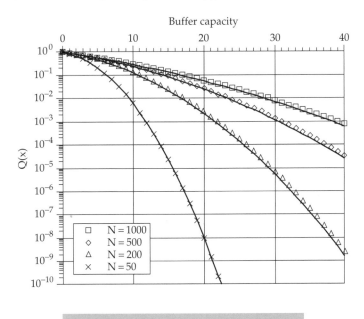

Figure 8.5. Comparison of Exact and Approximate Results for N·D/D/1 at a Load of 95%, and the Mathcad Code to Generate (x, y) Values for Plotting the Graph

[8.2] helps to identify explicitly the effect of the parameters:

$$Q(x) = e^{-2x\left(\frac{x}{N}+\frac{1-\rho}{\rho}\right)}$$

Figure 8.5 shows how the approximation compares with exact results from the N·D/D/1 analysis for a load of 95%. The approximate results are shown as lines, and the exact results as markers. In this case the approximation is in very good agreement. Figure 8.6 shows how the

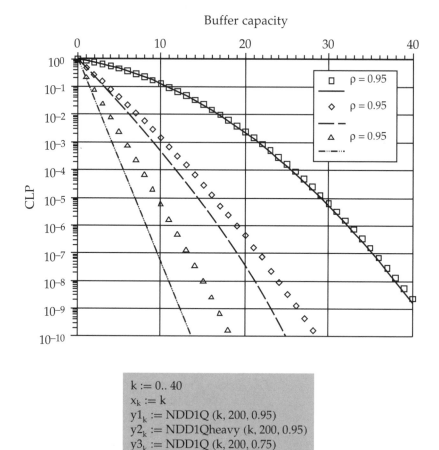

Figure 8.6. Comparison of Exact and Approximate Results for N·D/D/1 for a variety of Loads, with $N = 200$, and the Mathcad Code to Generate (x, y) Values for Plotting the Graph

approximation compares for three different loads. For low utilizations, the approximate method underestimates the cell loss.

Note that the form of the equation is similar to the approximation for the M/D/1 queue, with the addition of a quadratic term in x, the queue size. So, for small values of x, N·D/D/1 queues behave in a manner similar to M/D/1 queues with the same utilization. But for larger values of x the quadratic term dominates; this reduces the probability of larger queues occurring in the N·D/D/1, compared to the same size queue in the M/D/1 system. Thus we can see how the Poisson process is a useful approximation for N CBR sources, particularly for large N: as $N \rightarrow \infty$, the quadratic term disappears and the heavy traffic approximation to the N·D/D/1 becomes the same as that for the M/D/1. In Chapter 14 we revisit the M/D/1 to develop a more accurate formula for the overflow probability that both complements and extends the analysis presented in this chapter (see also [8.3]).

CELL-SCALE QUEUEING IN SWITCHES

It is important not to assume that cell-scale queueing arises only as a result of source multiplexing. If we now take a look at switching, we will find that the same effect arises. Consider the simple output buffered 2 × 2 switching element shown in Figure 8.7.

Here we can see a situation analogous to that of multiplexing the CBR sources. Both of the input ports into the switch carry cells coming from any number of previously multiplexed sources. Figure 8.8 shows a typical scenario; the cell streams on the input to the switching element are the output of another buffer, closer to the sources. The same queueing principle applies at the switch output buffer as at the source multiplexor: the sources may all be CBR, and the individual input ports to the switch may contain cells such that their aggregate arrival rate is less than the

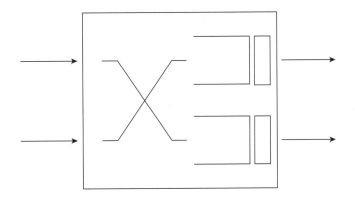

Figure 8.7.   An Output Buffered 2 × 2 Switching Element

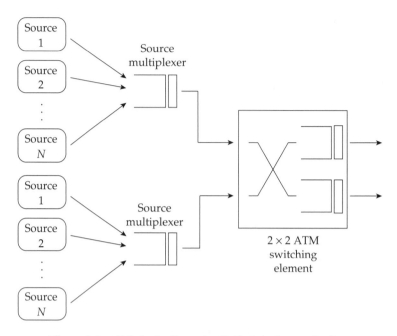

Figure 8.8. Cell-Scale Queueing in Switch Output Buffers

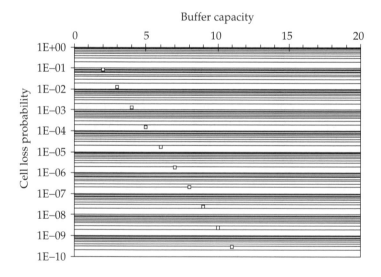

Figure 8.9. Cell Loss at the Switch Output Buffer

output rate of either of the switch output ports, *but still there can be cell loss in the switch*. Figure 8.9 shows an example of the cell loss probabilities for either of the output buffers in the switch for the scenario illustrated in Figure 8.8. This assumes that the output from each source multiplexor is a Bernoulli process, with parameter $p' = 0.5$, and that the cells are routed in equal proportions to the output buffers of the switching element. Thus the cell-scale queueing in each of the output buffers can be modelled with binomial input, where $M = 2$ and $p = 0.25$.

So, even if the whole of the ATM network is dedicated to carrying only CBR traffic, there is a need for buffers in the switches to cope with the cell-scale queueing behaviour. This is inherent to ATM; it applies even if the network allocates the peak rate to *variable*-bit-rate sources. Buffering is required, because multiple streams of cells are multiplexed together. It is worth noting, however, that the cell-scale queueing effect (measured by the CLP against the buffer capacity) falls away very rapidly with increasing buffer length – so we only need short buffers to cope with it, and to provide a cell loss performance in accord with traffic requirements. This is not the case with the burst-scale queueing behaviour, as we will see in Chapter 9.

9 Burst-Scale Queueing

information overload!

ATM QUEUEING BEHAVIOUR

We have seen in the previous chapter that queueing occurs with CBR traffic when two or more cells arrive during a time slot. If a particular source is CBR, we know that the next cell from it is going to arrive after a fixed duration given by the period, D, of the source, and this gives the ATM buffer some time to recover from multiple arrivals in any time slot when a number of sources are multiplexed together (hence the result that Poisson arrivals are a worst-case model for cell-scale queueing).

Consider the arrivals from all the CBR sources as a rate of flow of cells. Over the time interval of a single slot, the input rate varies in integer multiples of the cell slot rate (353 208 cell/s) according to the number of arrivals in the slot. But that input rate is very likely to change to a different value at the next cell slot; and the value will often be zero. It makes more sense to define the input rate in terms of the cycle time, D, of the CBR sources, i.e. 353 208/D cell/s. For the buffer to be able to recover from multiple arrivals in a slot, the number of CBR sources, N, must be less than the inter-arrival time D, so the total input rate 353 208 $\cdot N/D$ cell/s is less than the cell slot rate.

Cell-scale queueing analysis quantifies the effect of having simultaneous arrivals according to the relative phasing of the CBR streams, so we define simultaneity as being within the period of one cell slot.

Let's relax our definition of simultaneity, so that the time duration is a number of cell slots, somewhat larger than one. We will also alter our definition of an arrival from a single source; no longer is it a single cell, but a *burst* of cells during the defined period. Queueing occurs when the total number of cells arriving from simultaneous (or overlapping) bursts exceeds the number of cell slots in that 'simultaneous' period.

But how do we define the length of the 'simultaneous' period? Well, we don't: we define the source traffic using cell rates, and assume that these rates are on for long enough such that each source contributes rather more than one cell. Originally we considered CBR source traffic, whose behaviour was characterized by a fixed-length inactive state followed by the arrival of a single cell. For variable bit-rate (VBR), we redefine this behaviour as a long inactive state followed by an active state producing a burst of cells (where 'burst' is defined as a cell arrival rate over a period of time). The state-based sources in Chapter 6 are examples of models for VBR traffic.

With these definitions the condition for queueing is that the *total input rate of simultaneous bursts must exceed the cell slot rate of the ATM buffer*. This is called 'burst-scale queueing'. For the N CBR sources there is no burst-scale queueing because the total input rate of the simultaneous and continuous bursts of rate $353\,208/D$ cell/s is less than the cell slot rate.

Let's take a specific example, as shown in Figure 9.1. Here we have two VCs with fixed rates of 50% and 25% of the cell slot rate. In the first 12 time slots, the cells of the 25% VC do not coincide with those of the 50% VC and every cell can enter service immediately (for simplicity, we show this as happening in the same slot). In the second set of 12 time slots, the cells of the 25% VC do arrive at the same time as some of those in the 50% VC, and so some cells have to wait before being served. This is cell-scale queueing; the number of cells waiting is shown in the graph.

Now, let's add in a third VC with a rate of 33% of the cell slot rate (Figure 9.2). The total rate exceeds the queue service rate and over a period of time the number of cells waiting builds up: in this case there are two more arrivals than available service slots over the period shown in

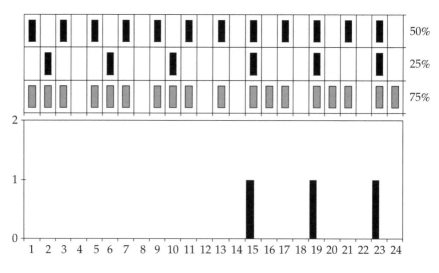

Figure 9.1. Cell Scale Queueing Behaviour

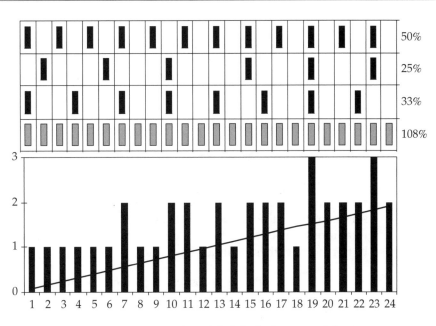

Figure 9.2. Burst-Scale and Cell-Scale Queueing Behaviour

the diagram. This longer-term queueing is the burst-scale queueing and is shown as a solid line in the graph. There is still the short-term cell-scale queueing, represented by the fluctuations in the number in the queue. ATM queueing comprises both types of behaviour.

BURST-SCALE QUEUEING BEHAVIOUR

The previous example showed that an input rate exceeding the service capacity by 8%, i.e. by 0.08 cells per time slot, would build up over a period of 24 time slots to a queue size of $0.08 \times 24 \approx 2$ cells. During this period (of about 68 µs) there were 26 arriving cells, but would be only 24 time slots in which to serve them: i.e. an excess of 2 cells. These two cells are called 'excess-rate' cells because they arise from 'excess-rate' bursts. Typical bursts can last for durations of milliseconds, rather than microseconds. So, in our example, if the excess rate lasts for 2400 time slots (6.8 ms) then there would be about 200 excess-rate cells that must be held in a buffer, or lost.

We can now distinguish between buffer storage requirements for cell-scale queueing (of the order of tens of cells) and for burst-scale queueing (of the order of hundreds of cells). Of course, there is only one buffer, through which all the cells must pass: what we are doing is identifying the two components of demand for temporary storage space. Burst-scale queueing analyses the demand for the temporary storage of these excess-rate cells.

We can identify two parts to this excess-rate demand, and analyse the parts separately. First, what is the probability that an arriving cell is an excess-rate cell? This is the same as saying that the cell needs burst-scale buffer storage. Then, secondly, what is the probability that such a cell is lost, i.e. the probability that a cell is lost, given that it is an excess-rate cell? We can then calculate the overall cell loss probability arising from burst-scale queueing as:

$$\Pr\{\text{cell is lost}\} \approx \Pr\{\text{cell is lost|cell needs buffer}\} \cdot \Pr\{\text{cell needs buffer}\}$$

The probability that a cell needs the buffer is called the burst-scale loss factor; this is found by considering how the input rate compares with the service rate of the queue. A cell needs to be stored in the buffer if the total input rate exceeds the queue's service rate. If there is no burst-scale buffer storage, these cells are lost, and

$$\Pr\{\text{cell is lost}\} \approx \Pr\{\text{cell needs buffer}\}$$

The probability that a cell is lost given that it needs the buffer is called the 'burst-scale delay factor'; this is the probability that an *excess-rate* cell is lost. If the burst-scale buffer size is 0, then this probability is 1, i.e. all excess-rate cells are lost. However, if there is some buffer storage, then only some of the excess-rate cells will be lost (when this buffer storage is full).

Figure 9.3 shows how these two factors combine on a graph of cell loss probability against the buffer capacity. The burst-scale delay factor is shown as a straight line with the cell loss decreasing as the buffer capacity increases. The burst-scale loss factor is the intersection of the straight line with the zero buffer axis.

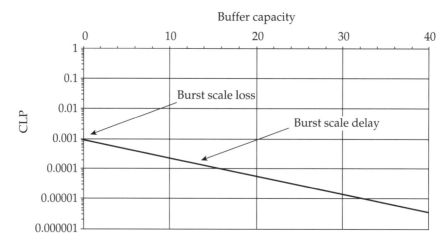

Figure 9.3. The Two Factors of Burst-Scale Queueing Behaviour

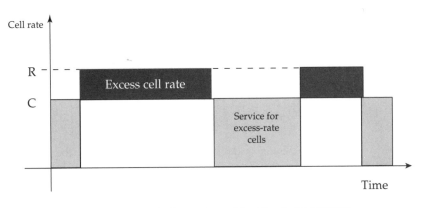

Figure 9.4.   Burst-Scale Queueing with a Single ON/OFF Source

FLUID-FLOW ANALYSIS OF A SINGLE SOURCE – PER-VC QUEUEING

The simplest of all burst-scale models is the single ON/OFF source feeding an ATM buffer. When the source is ON, it produces cells at a rate, R, overloading the service capacity, C, and causing burst-scale queueing; when OFF, the source sends no cells, and the buffer can recover from this queueing by serving excess-rate cells (Figure 9.4). In this very simple case, there is no cell-scale queueing because only one source is present. This situation is essentially that of per-VC queueing: an output port is divided into multiple virtual buffers, each being allocated a share of the service capacity and buffer space available at the output port. Thus, in the following analysis, C can be thought of as the *share* of service capacity allocated to a virtual buffer for this particular VC connection. We revisit this in Chapter 16 when we consider per-flow queueing in IP.

There are two main approaches to this analysis. The historical approach is to model the flow of cells into the buffer as though it were a continuous fluid; this ignores the structure of the flow (e.g. bits, octets, or cells). The alternative is the discrete approach, which actually models the individual excess-rate cells.

CONTINUOUS FLUID-FLOW APPROACH

The source model for this approach was summarized in Chapter 6; the state durations are assumed to be exponentially distributed. A diagram of the system is shown in Figure 9.5. Analysis requires the use of partial differential equations and the derivation is rather too complex in detail to merit inclusion here (see [9.1] for details). However, the equation for the excess-rate loss probability is

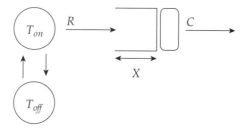

Figure 9.5. Source Model and Buffer Diagram for the Continuous Fluid-Flow Analysis

$$\text{CLP}_{\text{excess-rate}} = \frac{(C - \alpha \cdot R) \cdot e^{\left(\frac{-X \cdot (C - \alpha \cdot R)}{T_{on} \cdot (1 - \alpha) \cdot (R - C) \cdot C}\right)}}{(1 - \alpha) \cdot C - \alpha \cdot (R - C) \cdot e^{\left(\frac{-X \cdot (C - \alpha \cdot R)}{T_{on} \cdot (1 - \alpha) \cdot (R - C) \cdot C}\right)}}$$

where

R = ON rate
C = service rate of queue
X = buffer capacity of queue
T_{on} = mean duration in ON state
T_{off} = mean duration in OFF state

and

$$\alpha = \frac{T_{on}}{T_{on} + T_{off}} = \text{probability that the source is active}$$

Note that $\text{CLP}_{\text{excess-rate}}$ is the probability that a cell is lost given that it is an excess-rate cell. The probability that a cell is an excess-rate cell is simply the proportion of excess-rate cells to all arriving cells, i.e. $(R - C)/R$. Thus the overall cell loss probability is

$$\text{CLP} = \frac{R - C}{R} \cdot \text{CLP}_{\text{excess-rate}}$$

Another way of looking at this is to consider the number of cells lost in a time period, T.

$$R \cdot \alpha \cdot T \cdot \text{CLP} = (R - C) \cdot \alpha \cdot T \cdot \text{CLP}_{\text{excess-rate}}$$

The mean number of cells arriving in one ON/OFF cycle is $R \cdot T_{on}$, so the mean arrival rate is simply $R \cdot \alpha$. The mean number of cells arriving during the time period is $R \cdot \alpha \cdot T$. Thus the number of cells actually lost (on average) during time period T is given by the left-hand side of the equation. But cells are only lost when the source is in the ON state, i.e. when there are excess-rate arrivals. Thus the mean number of excess-rate arrivals in one ON/OFF cycle is $(R - C) \cdot T_{on}$, and so the mean excess rate is simply $(R - C) \cdot \alpha$. The number of excess-rate cells arriving during the time period is $(R - C) \cdot \alpha \cdot T$, and so the number of excess-rate cells actually lost during a time period T is given by the right-hand side of the equation. There is no other way of losing cells, so the two sides of the equation are indeed equal, and the result for CLP follows directly.

We will take an example and put numbers into the formula later on, when we can compare with the results for the discrete approach.

DISCRETE 'FLUID-FLOW' APPROACH

This form of analysis 'sees' each of the excess-rate arrivals [9.2]. The derivation is simpler than that for the continuous case, and the approach to deriving the balance equations is a useful alternative to that described in Chapter 7. Instead of finding the state probabilities at the end of a time slot, we find the probability that an arriving excess-rate cell finds k cells in the buffer. If an arriving excess-rate cell finds the buffer full, it is lost, and so $CLP_{excess-rate}$ is simply the probability of this event occurring.

We start with the same system model and parameters as for the continuous case, shown in Figure 9.5. The system operation is as follows:

IF the source is in the OFF state AND

a)   the buffer is empty, THEN it remains empty
b)   the buffer is not empty, THEN it empties at a constant rate C

IF the source is in the ON state AND

a)   the buffer is not full, THEN it fills at a constant rate $R - C$
b)   the buffer is full, THEN cells are lost at a constant rate $R - C$

As was discussed in Chapter 6, in the source's OFF state no cells are generated, and the OFF period lasts for a geometrically distributed number of time slots. In the ON state, cells are generated at a rate of R. But for this analysis we are only interested in the excess-rate arrivals, so in the ON state we say that excess-rate cells are generated at a rate of $R - C$ and the ON period lasts for a geometrically distributed number of excess-rate arrivals. In each state there is a Bernoulli process: in the OFF state, the probability of being silent for another time slot is s; in the ON state, the probability of generating another excess-rate arrival is a. The model is shown in Figure 9.6.

Once the source has entered the OFF state, it remains there for at least one time slot; after each time slot in the OFF state the source remains in the OFF state with probability s, or enters the ON state with probability $1 - s$.

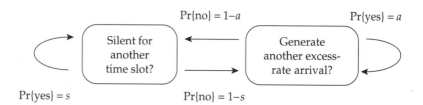

Figure 9.6. The ON/OFF Source Model for the Discrete 'Fluid-Flow' Approach

On entry into the ON state, the model generates an excess-rate arrival; after each arrival the source remains in the ON state and generates another arrival with probability a, or enters the OFF state with probability $1 - a$. This process of arrivals and time slots is shown in Figure 9.7.

Now we need to find a and s in terms of the system parameters, R, C, T_{on} and T_{off}. From the geometric process we know that the mean number of excess-rate cells in an ON period is given by

$$E[on] = \frac{1}{1 - a}$$

But this is simply the mean duration in the ON state multiplied by the excess rate, so

$$E[on] = \frac{1}{1 - a} = T_{on} \cdot (R - C)$$

giving

$$a = 1 - \frac{1}{T_{on} \cdot (R - C)}$$

In a similar manner, the mean number of empty time slots in the OFF state is

$$E[off] = \frac{1}{1 - s} = T_{off} \cdot C$$

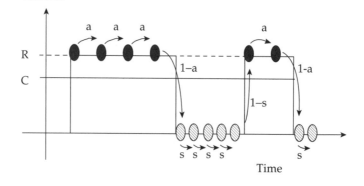

Figure 9.7. The Process of Arrivals and Time Slots for the ON/OFF Source Model

giving

$$s = 1 - \frac{1}{T_{off} \cdot C}$$

In Chapter 7, we developed balance equations that related the state of the buffer at the end of the current time slot with its state at the end of the previous time slot. This required knowledge of all the possible previous states and how the presence or absence of arrivals could achieve a transition to the current state. For this discrete fluid-flow approach, we use a slightly different form of balance equation, developed according to the so-called 'line crossing' method.

Consider the contents of a queue varying over time, as shown in Figure 9.8. If we 'draw a line' between states of the queue (in the figure we have drawn one between state ⟨there are 3 in the queue⟩ and state ⟨there are 4 in the queue⟩) then for every up-crossing through this line, there will also be a down-crossing (otherwise the queue contents would increase for ever). Since we know that a probability value can be represented as a proportion, we can equate the proportion of transitions that cause the queue to cross up through the line (probability of crossing up) with the proportion of transitions that cause it to cross down through the line (probability of crossing down). This will work for a line drawn through any adjacent pair of states of the queue.

We define the state probability as

$p(k) = \Pr\{$an arriving excess-rate cell finds k cells in the buffer$\}$

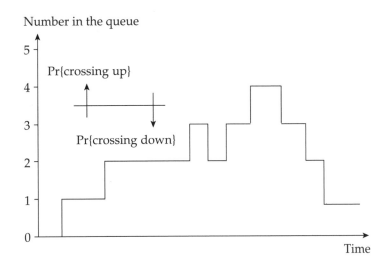

Figure 9.8. The Line Crossing Method

An excess-rate cell which arrives to find X cells in the buffer, where X is the buffer capacity, is lost, so

$$\mathrm{CLP}_{\text{excess-rate}} = p(X)$$

The analysis begins by considering the line between states X and $X - 1$. This is shown in Figure 9.9.

Since we are concerned with the state that an arriving excess-rate cell sees, we must consider arrivals one at a time. Thus the state can only ever increase by one. This happens when an arriving excess-rate cell sees $X - 1$ in the queue, taking the queue state up to X, and another excess-rate cell follows immediately (without any intervening empty time slots) to see the queue in state X. So, the probability of going up is

$$\Pr\{\text{going up}\} = a \cdot p(X - 1)$$

To go down, an arriving excess-rate cell sees X in the queue and is lost (because the queue is full), and then there is a gap of at least one empty time slot, so that the next arrival sees fewer than X in the queue. (If there is no gap, then the queue will remain full and the next arrival will see X as well.) So, the probability of going down is

$$\Pr\{\text{going down}\} = (1 - a) \cdot p(X)$$

Equating the probabilities of going up and down, and rearranging gives

$$p(X - 1) = \frac{1 - a}{a} \cdot p(X)$$

We can do the same for a line between states $X - 1$ and $X - 2$. Equating probabilities gives

$$a \cdot p(X - 2) = (1 - a) \cdot s \cdot p(X) + (1 - a) \cdot s \cdot p(X - 1)$$

The left-hand side is the probability of going up, and is essentially the same as before. The probability of going down, on the right-hand side of the equation, contains two possibilities. The first term is for an arriving excess-rate cell which sees X in the queue and is lost (because the queue

$$a{\cdot}p(X{-}1) \uparrow \quad \frac{X}{X - 1} \quad \downarrow (1{-}a){\cdot}p(X)$$

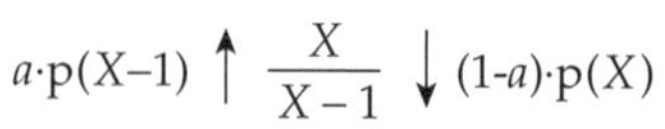

Figure 9.9. Equating Up- and Down-Crossing Probabilities between States X and $X - 1$

is full), and then there is a gap of at least two empty time slots, so that the next arrival sees fewer than $X - 1$ in the queue. The second term is for an arriving excess-rate cell which sees $X - 1$ in the queue, taking the state of the queue up to X, and then there is a gap of at least two empty time slots, so that the next arrival sees fewer than $X - 1$ in the queue. Rearranging, and substituting for $p(X)$, gives

$$p(X - 2) = \frac{s}{a} \cdot p(X - 1)$$

In the general case, for a line between $X - i + 1$ and $X - i$, the probability of going up remains the same as before, i.e. the only way to go up is for an arrival to see $X - i$, and to be followed immediately by another arrival which sees $X - i + 1$. The probability of going down consists of many components, one for each state above $X - i$, but they can be arranged in two groups: the probability of coming down from $X - i + 1$ itself; and the probability of coming down to below $X - i + 1$ from above $X - i + 1$. This latter is just the probability of going down between $X - i + 2$ and $X - i + 1$ multiplied by s, which is the same as going up from $X - i + 1$ multiplied by s. This is illustrated in Figure 9.10.

The general equation then is

$$p(X - i) = \frac{s}{a} \cdot p(X - i + 1)$$

The state probabilities form a geometric progression, which can be expressed in terms of $p(X)$, a and s, for $i > 0$:

$$p(X - i) = \left(\frac{s}{a}\right)^i \cdot \frac{1 - a}{s} \cdot p(X)$$

The probabilities must sum to 1, so

$$\sum_{i=0}^{X} p(X - i) = p(X) + \sum_{i=1}^{X} \left(\frac{s}{a}\right)^i \cdot \frac{1 - a}{s} \cdot p(X) = 1$$

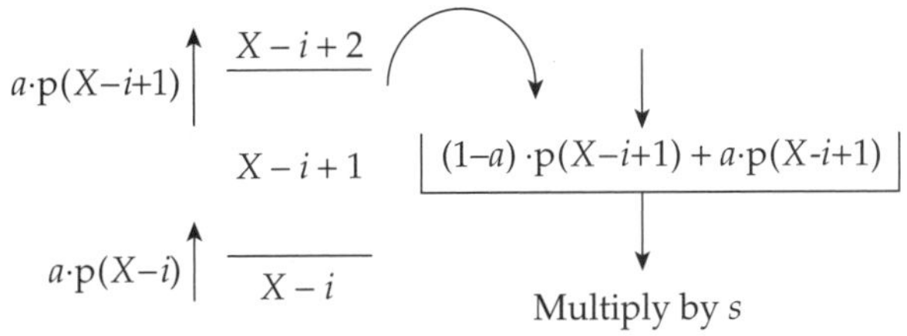

Figure 9.10. Equating Up- and Down-Crossing Probabilities in the General Case

which can be rearranged to give the probability that an excess-rate arrival sees a full queue, i.e. the excess-rate cell loss probability

$$p(X) = \frac{1}{1 + \sum\limits_{i=1}^{X} \left(\frac{s}{a}\right)^{i} \cdot \frac{1-a}{s}}$$

This can be rewritten as

$$p(X) = \frac{1}{1 + \left(\left(\frac{s}{a}\right)^{X} - 1\right) \cdot \left(\frac{1-a}{s-a}\right)}$$

which is valid except when $a = s$ (in which case the previous formula must be used). As in the case of the continuous fluid-flow analysis, the overall cell loss probability is given by

$$\text{CLP} = \frac{R-C}{R} \cdot \text{CLP}_{\text{excess-rate}} = \frac{R-C}{R} \cdot p(X)$$

COMPARING THE DISCRETE AND CONTINUOUS FLUID-FLOW APPROACHES

Let's use the practical example of silence-suppressed telephony, with the following parameter values:

$$R = 167 \text{ cell/s}$$

$$T_{on} = 0.96 \text{ seconds}$$

$$T_{off} = 1.69 \text{ seconds}$$

Thus

$$\alpha = \frac{0.96}{0.96 + 1.69} = 0.362$$

and the mean arrival rate

$$\lambda = \alpha \cdot R = 60.5 \text{ cell/s}$$

In order to have burst-scale queueing, the service capacity, C, must be less than the cell rate in the active state, R. Obviously this does not correspond to a normal ATM buffer operating at 353 208 cell/s (although it does if we are considering a virtual buffer with per-VC queueing). We will see, also, that one application of this analysis is in connection admission control for estimating a bandwidth value, C, to allocate to a source in order to meet a cell loss probability requirement. Figure 9.11 shows the overall

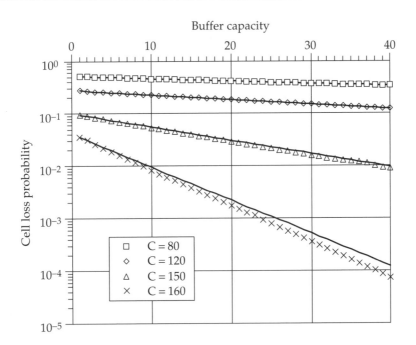

$$k := 1.. \quad 40$$

$$\text{CLPcontFF}(C, \alpha, R, X, \text{Ton}) := \left(\frac{R-C}{R}\right) \cdot \frac{(C-\alpha \cdot R) \cdot e^{\left[\frac{-X \cdot (C-\alpha \cdot R)}{\text{Ton} \cdot (1-\alpha) \cdot (R-C) \cdot C}\right]}}{(1-\alpha) \cdot C - \alpha \cdot (R-C) \cdot e^{\left[\frac{-X \cdot (C-\alpha \cdot R)}{\text{Ton} \cdot (1-\alpha) \cdot (R-C) \cdot C}\right]}}$$

$$\text{CLPdiscFF}(C, \alpha, R, X, \text{Ton}) := \left| \begin{array}{l} \text{Toff} \leftarrow \text{Ton} \cdot \dfrac{1-\alpha}{\alpha} \\[2ex] a \leftarrow 1 - \dfrac{1}{\text{Ton} \cdot (R-C)} \\[2ex] s \leftarrow 1 - \dfrac{1}{\text{Toff} \cdot C} \\[2ex] \dfrac{R-C}{R} \cdot \dfrac{1}{1 + \left[\left(\dfrac{s}{a}\right)^X - 1\right] \cdot \dfrac{1-a}{s-a}} \end{array} \right.$$

$X_k := k$

$y1_k := \text{CLPdiscFF}(80, 0.362, 167, k, 0.96)$

$y2_k := \text{CLPdiscFF}(120, 0.362, 167, k, 0.96)$

$y3_k := \text{CLPdiscFF}(150, 0.362, 167, k, 0.96)$

$y4_k := \text{CLPdiscFF}(160, 0.362, 167, k, 0.96)$

$y5_k := \text{CLPcontFF}(80, 0.362, 167, k, 0.96)$

$y6_k := \text{CLPcontFF}(120, 0.362, 167, k, 0.96)$

$y7_k := \text{CLPcontFF}(150, 0.362, 167, k, 0.96)$

$y8_k := \text{CLPcontFF}(160, 0.362, 167, k, 0.96)$

Figure 9.11.   Cell Loss Probability against Buffer Capacity for a Single ON/OFF Source

Table 9.1. Average Number of Excess-Rate Cells in an Active Period

Capacity (cell/s)	Average number of excess-rate cells per active state
80	83.52
120	45.12
150	16.32
160	6.72

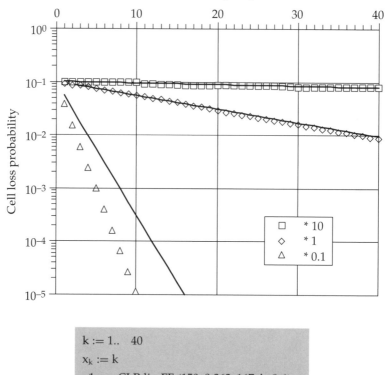

$$k := 1.. \ 40$$
$$x_k := k$$
$$y1_k := \text{CLPdiscFF} \ (150, 0.362, 167, k, 9.6)$$
$$y2_k := \text{CLPdiscFF} \ (150, 0.362, 167, k, 0.96)$$
$$y3_k := \text{CLPdiscFF} \ (150, 0.362, 167, k, 0.096)$$
$$y4_k := \text{CLPcontFF} \ (150, 0.362, 167, k, 9.6)$$
$$y5_k := \text{CLPcontFF} \ (150, 0.362, 167, k, 0.96)$$
$$y6_k := \text{CLPcontFF} \ (150, 0.362, 167, k, 0.096)$$

Figure 9.12. The Effect of Scaling the Mean State Durations, T_{on} and T_{off}, when $C = 150$ cell/s

cell loss probability plotted against the buffer capacity, X, as the service capacity is varied between the mean and peak cell rates of the source. The discrete fluid-flow results are shown as markers, and the continuous as solid lines.

As the service capacity gets closer to the ON rate, R, the gradient steepens. This means that the buffer is better able to cope with the bursts of excess-rate cells. We can see more clearly why this is so by looking at Table 9.1, which shows the average number of excess-rate cells in an active period. When this number is large relative to the capacity of the buffer, then the buffer does not cope very well because it only takes a fraction of an average burst to fill it up. It would not make much difference if there was no buffer space–there is so little difference to the cell loss over the range of buffer capacity shown. The buffer only makes a difference to the cell loss if the average excess-rate burst length is less than the buffer capacity, i.e. when it would take a number of bursts to fill the buffer. Notice that it is only in these circumstances that the discrete and continuous fluid-flow results show any difference; and then the discrete approach is more accurate because it does not include the 'fractions' of cells allowed by the continuous fluid-flow analysis. These small amounts actually represent quite a large proportion of an excess-rate burst when the average number of excess-rate cells in the burst is small.

Figure 9.12 illustrates the strong influence of the average state durations on the results for cell loss probability. Here, $C = 150$ cell/s, with other parameter values as before, and the T_{on} and T_{off} values have been scaled by 0.1, 1 and 10. In each case the load on the buffer remains constant at a value of

$$\alpha \cdot \frac{R}{C} = 0.362 \cdot \frac{167}{150} = 0.403$$

MULTIPLE ON/OFF SOURCES OF THE SAME TYPE

Let's now consider burst-scale queueing when there are multiple ON/OFF sources being fed through an ATM buffer. Figure 9.13 shows a diagram of the system, with the relevant source and buffer parameters. There are N identical sources, each operating independently, sending cells into an ATM buffer of service capacity C cell/s and finite size X cells. The average ON and OFF durations are denoted T_{on} and T_{off}, as before; the cell rate in the active state is h cell/s, so the mean cell rate for each source is

$$m = h \cdot \frac{T_{on}}{T_{on} + T_{off}}$$

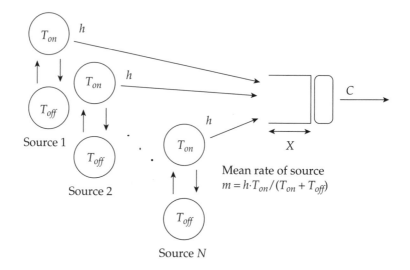

Figure 9.13. Multiple ON/OFF Sources Feeding an ATM Buffer

and the probability that the source is active is

$$\alpha = \frac{m}{h} = \frac{T_{on}}{T_{on} + T_{off}}$$

which is also called the 'activity factor'.

The condition for burst-scale queueing is that the total input rate from active sources must be greater than the service rate of the buffer. An important parameter, then, is how many times the peak rate, h, fits into the service capacity, C, denoted by N_0:

$$N_0 = \frac{C}{h}$$

This may well not be an integer value. If we round the value up, to $\lceil N_0 \rceil$ (this notation means 'take the first integer above N_0'), this gives the minimum number of sources required for burst-scale queueing to take place. If we round the value down, to $\lfloor N_0 \rfloor$ (this notation means 'take the first integer below N_0'), this gives the maximum number of sources we can have in the system *without* having burst-scale queueing.

We saw earlier in the chapter that the burst-scale queueing behaviour can be separated into two components: the burst-scale loss factor, which is the probability that a cell is an excess-rate cell; and the burst-scale delay factor, which is the probability that a cell is lost given that it is an excess-rate cell. Both factors contribute to quantifying the cell loss: the burst-scale loss factor gives the cell loss probability if we assume there is no buffer;

this value is multiplied by the burst-scale delay factor to give the cell loss probability if we assume there *is* a buffer of some finite capacity.

THE BUFFERLESS APPROACH

For multiple ON/OFF sources, we start by assuming there is no buffer and calculate the burst-scale loss factor. For the single source, this is simply the proportion of cells that are excess-rate cells, i.e. $(R - C)/R$, or with the new parameters, $(h - C)/h$. Another way of looking at this is the *mean* excess rate divided by the *mean* arrival rate:

$$\Pr\{\text{cell needs buffer}\} = \frac{\alpha \cdot (h - C)}{m} = \frac{\alpha \cdot (h - C)}{\alpha \cdot h} = \frac{h - C}{h}$$

The probability of an excess rate of $h - C$ is the same as the probability that the source is active, i.e. α, hence the mean excess rate is just $\alpha \cdot (h - C)$. In the case of multiple sources, we need to calculate the probability that n sources are active, where $N_0 < n \leqslant N$ and multiply by the excess rate, $n \cdot h - C$. This probability is given by the binomial distribution

$$p_n = \frac{N!}{n! \cdot (N - n)!} \cdot \alpha^n \cdot (1 - \alpha)^{N-n}$$

and so the mean excess rate is

$$\sum_{n=\lceil N_0 \rceil}^{N} p_n \cdot (n \cdot h - C)$$

The mean arrival rate is simply $N \cdot m$, so the probability that a cell needs the buffer is given by the ratio of the mean excess rate to the mean arrival rate:

$$\Pr\{\text{cell needs buffer}\} = \frac{\displaystyle\sum_{n=\lceil N_0 \rceil}^{N} p_n \cdot (n \cdot h - C)}{N \cdot m}$$

which, if we substitute for $C = N_0 \cdot h$ and $\alpha = m/h$, gives

$$\Pr\{\text{cell needs buffer}\} = \frac{\displaystyle\sum_{n=\lceil N_0 \rceil}^{N} p_n \cdot (n - N_0)}{N \cdot \alpha}$$

Let's put some numbers into this formula, using the example of two different types of video source, each with a mean bit-rate of 768 kbit/s and peak bit-rates of either 4.608 Mbit/s or 9.216 Mbit/s. The corresponding

Table 9.2. Parameter Values

h (cell/s)	$\alpha = \dfrac{2000}{h}$	$N_0 = \dfrac{353\,207.55}{h}$
12 000	0.167	29.43
24 000	0.083	14.72

cell rates are $m = 2000$ cell/s and $h = 12\,000$ cell/s or $24\,000$ cell/s, and the other parameter values are shown in Table 9.2.

Figure 9.14 shows how the probability that a cell needs the buffer increases with the number of video sources being multiplexed through the buffer. The minimum number of sources needed to produce burst-scale queueing is 30 (for $h = 12\,000$) or 15 (for $h = 24\,000$). The results show that about twice these values (60 and 30, respectively) produce 'loss' probabilities of about 10^{-10}, increasing to between 10^{-1} and 10^{-2} for 150 of either source (see Figure 9.14). For both types of source the mean rate, m, is 2000 cell/s, so the average load offered to the buffer, as a fraction of its service capacity, ranges from $30 \times 2000/353\,208 \approx 17\%$ up to $150 \times 2000/353\,208 \approx 85\%$.

We know from Chapter 6 that the binomial distribution can be approximated by the Poisson distribution when the number of sources, N, becomes large. This can be used to provide an approximate result for Pr{cell needs buffer}, the burst-scale loss factor, and it has the advantage of being less demanding computationally because there is no summation [9.3].

$$\text{Pr\{cell needs buffer\}} \approx \frac{1}{(1 - \rho)^2 \cdot N_0} \cdot \frac{(\rho \cdot N_0)^{\lfloor N_0 \rfloor}}{\lfloor N_0 \rfloor!} \cdot e^{-\rho \cdot N_0}$$

where the offered load, ρ, is given by

$$\rho = \frac{N \cdot m}{C} = N \cdot \frac{m}{h} \cdot \frac{h}{C} = \frac{N \cdot \alpha}{N_0}$$

Figure 9.15 shows results for ON/OFF sources with peak rate $h = 12\,000$ cell/s, and mean rates varying from $m = 2000$ cell/s ($\alpha = 0.167$) down to 500 cell/s ($\alpha = 0.042$). N_0 is fixed at 29.43, and the graph plots the 'loss' probability varying with the offered load, ρ. We can see that for any particular value of ρ the burst-scale loss factor increases, as the activity factor, α, decreases, towards an upper limit given by the approximate result. The approximation thus gives a conservative estimate of the probability that a cell needs the buffer. Note that as the activity factor decreases, the number of sources must increase to maintain the constant load, taking it into the region for which the Poisson approximation is valid.

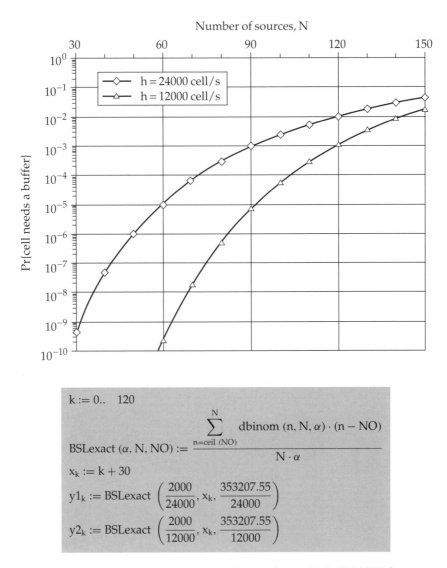

Figure 9.14. The Bufferless Approach–Results for Multiple ON/OFF Sources

How does this Poisson approximation change our view of the source process? Instead of considering N identical ON/OFF sources each with probability α of being in the active state and producing a burst of fixed rate h, we are modelling the traffic as just one Poisson source which produces overlapping bursts. The approximation equates the average number of active sources with the average number of bursts in progress. It's similar to our definition of traffic intensity, but at the burst level.

The average number of active sources is simply $N \cdot \alpha$; now, recalling that the probability of being active is related to the average durations in

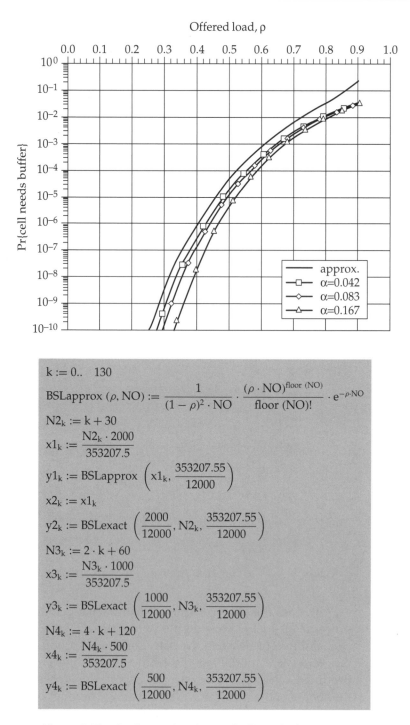

Figure 9.15. An Approximation to the Burst-Scale Loss Factor

the ON and OFF states:

$$\alpha = \frac{T_{on}}{T_{on} + T_{off}}$$

we can substitute for α to obtain

$$\text{average number of active sources} = T_{on} \cdot \frac{N}{T_{on} + T_{off}} = T_{on} \cdot \lambda$$

which is the average burst duration multiplied by the burst rate, λ (each source produces one burst every cycle time, $T_{on} + T_{off}$). This is the average number of bursts in progress.

THE BURST-SCALE DELAY MODEL

We are now in a position to extend the burst-scale analysis to finding the probability that an excess-rate cell is lost given that it is an excess-rate cell. With the bufferless approach, this probability is 1; every excess-rate cell is lost because we assume there is no buffer in which to store it temporarily. Now we assume that there is a finite amount of buffer space, X, as shown in Figure 9.13.

We will view the N ON/OFF sources as a single Poisson source producing bursts of cell rate h and duration T_{on} at a rate of λ bursts per second, where

$$\lambda = \frac{N}{T_{on} + T_{off}}$$

Note that there is now no limit to the number of overlapping bursts; the Poisson model can exceed N simultaneous bursts. But if N is sufficiently large, the approximation to a Poisson source is reasonable. The average number of cells per burst, b, is given by:

$$b = T_{on} \cdot h$$

so the load offered to the queue, as a fraction of the service capacity is

$$\rho = \frac{b \cdot \lambda}{C}$$

If we substitute for b and λ (just to check) we obtain

$$\rho = \frac{T_{on} \cdot h \cdot \dfrac{N}{T_{on} + T_{off}}}{C} = \frac{N \cdot m}{C}$$

which is what we had for the bufferless model.

An approximate analysis of this burst-scale delay model uses the M/M/N queueing system (where the number of parallel servers, N, is taken to be the maximum number of bursts which can fit into the service capacity of the ATM buffer, N_0) to give the following estimate for the probability of loss [9.3]:

$$CLP_{\text{excess-rate}} = e^{-\left[N_0 \cdot \dfrac{X}{b} \cdot \dfrac{(1-\rho)^3}{4 \cdot \rho + 1}\right]}$$

This is similar in form to the heavy traffic approximations of Chapter 8: an exponential function of buffer capacity and utilization. Note that the buffer capacity here can be considered in units of the average burst length, i.e. as X/b.

Recall that for the N·D/D/1 approximation, N, the number of CBR sources, is in the denominator of the exponential function. With a constant load, as N increases, the gradient on the graph of cell loss against buffer capacity decreases, i.e. the buffer is less able to cope with more sources of smaller fixed cell rate. In contrast, this burst-scale delay result has N_0 in the numerator of the exponential function. N_0 is the minimum number of overlapping bursts required for burst-scale queueing. As N_0 increases, so does the gradient, and the buffer is better able to cope with more sources of smaller ON rates. Why? If it takes more bursts to achieve queueing in the buffer then the period of overlap will be smaller, reducing the effective size of the excess-rate burst. An intuitive way of viewing this is to *think* of b/N_0 as the average excess-rate burst length; then $N_0 \cdot X/b$ can be considered as the buffer capacity in units of the average excess-rate burst length.

Let's continue with the example of the video sources we used earlier. The mean cell rate for both types of source is $m = 2000$ cell/s and the peak cell rates are equal to either 12 000 cell/s or 24 000 cell/s. What we still need to specify are the state durations. If we assume that the ON state is equivalent to a highly active video frame, then we can use a value of 40 ms for T_{on}, which means the average number of cells per burst is $0.04 \times 12\,000 = 480$ cells or 960 cells respectively. T_{off} is given by

$$T_{off} = T_{on} \cdot \frac{h - m}{m}$$

so T_{off} takes values of 0.2 second or 0.44 second respectively. The ON/OFF source cycle times $(T_{on} + T_{off})$ are 0.24 s and 0.48 s, so the burst rates for the equivalent Poisson source of bursts are 4.167 (i.e. 1/0.24) or 2.083 times the number of sources, N, respectively.

Figure 9.16 shows the effect of the buffer capacity on the excess-rate cell loss when there are 60 sources, giving an offered load of 0.34. The results for three types of source are shown: the two just described, and

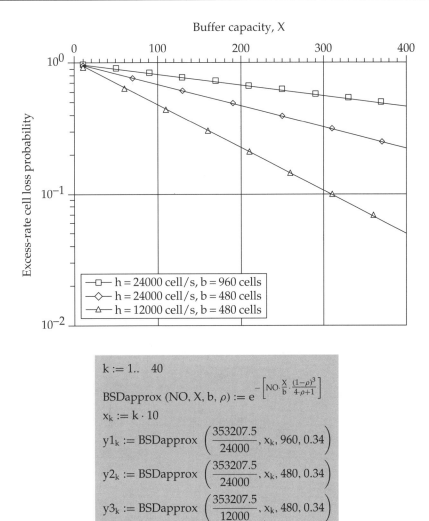

Figure 9.16. Probability of Excess-Rate Cell Loss

then the higher peak-rate source with an average active state duration of half the original. This makes the average burst length, b, the same as that for the lower-rate source. We can then make a fair assessment of the impact of N_0, with b and ρ kept constant. It is clear then that as the peak rate decreases, and therefore N_0 increases, the buffer is better able to cope with the excess-rate bursts.

Figure 9.17 shows how the two factors which make up the overall cell loss probability are combined. The buffer capacity value was set at 400 cells. This corresponds to a maximum waiting time of 1.1 ms. The burst-scale delay factor is shown for the two different peak rates as the

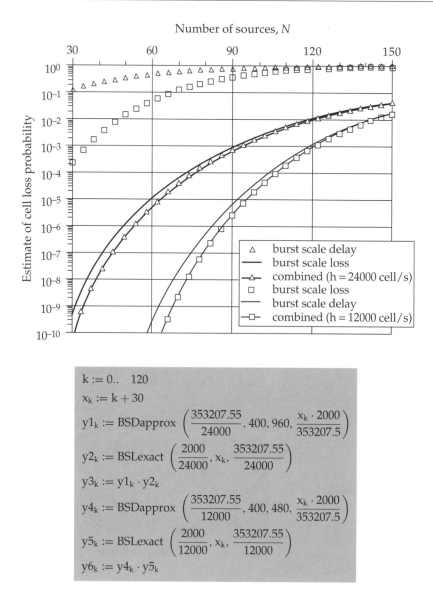

Figure 9.17. Combining Results for Burst-Scale Delay Factor with the Burst-Scale Loss Factor

curves with markers only. These results tend to an excess-rate loss probability of 1 as the number of sources, and hence the offered load increases. The burst-scale loss results from Figure 9.14 are shown as the lines without markers. The overall cell loss probabilities are the product of the two factors and are the results shown with both lines and markers. Notice that the extra benefit gained by having a large buffer for burst-scale queueing does not appear to be that significant, for the situation considered here.

10 Connection Admission Control

the net that likes to say YES!

No network operator likes to turn away business; if it does so too often customers are likely to take their business elsewhere. Yet if the operator always accepts any connection request, the network may become congested, unable to meet the negotiated performance objectives for the connections already established, with the likely outcome that many customers *will* take their business elsewhere.

Connection admission control (CAC) is the name for that mechanism which has to decide whether or not the bandwidth and performance requirements of a new connection can be supported by the network, in addition to those of the connections already established. If the new connection is accepted, then the bandwidth and performance requirements form a traffic contract between the user and the network. We have seen in Chapter 9 the impact that changes in traffic parameter values have on performance, whether it is the duration of a peak-rate burst, or the actual cell rate of a state. It is important then for the network to be able to ensure that the traffic does not exceed its negotiated parameter values. This is the function of usage parameter control. This in turn ensures that the network meets the performance requirements for all the connections it has admitted. Together, connection admission control and usage parameter control (UPC) are the main components in a traffic control framework which aims to prevent congestion occurring. Congestion is defined as a state of network elements (such as switching nodes and transmission links) in which the network is not able to meet the negotiated performance objectives. Note that congestion is to be distinguished from queue saturation, which may happen while still remaining within the negotiated performance objective.

In a digital circuit-switched telephone network the admission control problem is to find an unused circuit on a route from source to destination for a single type of traffic. If a 64 kbit/s circuit is not available, then the connection is blocked. In ATM the problem is rather more complicated: not only must the route be found, but also a check must be made at each link on a proposed route to ensure that the new connection, with whatever traffic characteristics, can be supported without violating the negotiated performance requirements of connections established over each link.

In this chapter we focus on how we may make the check on each link, by making use of the cell-scale and burst-scale queueing analysis of previous chapters.

THE TRAFFIC CONTRACT

How are the bandwidth and performance requirements of the traffic contract specified? In our burst-scale analysis so far, we have seen that there are three traffic parameters which are important in determining the type of queueing behaviour: peak cell rate, mean cell rate, and the average active state duration. For the performance requirement, we have concentrated on cell loss probability, but cell delay and CDV (cell-delay variation) can also be important, particularly for interactive services.

The number of bandwidth parameters in the traffic contract is closely related to the complexity of the CAC algorithm and the type of queueing behaviour that is being permitted on the network. The simplest approach is CAC based on peak cell rate only: this limits the combined peak cell rate of all VCs through a buffer to less than or equal to the service capacity of the buffer. In this case there is never any burst-scale queueing, so the CAC algorithm is based on cell-scale queueing analysis. The ITU Standards terminology for a traffic control framework based on peak cell rate only is 'deterministic bit-rate (DBR) transfer capability' [10.1]. The equivalent to this in ATM Forum terminology is 'constant bit-rate (CBR) service category' [10.2]. If we add another bandwidth parameter, the mean cell rate, to the traffic contract and allow the peak cell rate to exceed the service capacity, this is one form of what is called the 'statistical bit-rate (SBR) transfer capability'. In this case the CAC algorithm is based on both cell-scale queueing analysis and burst-scale loss factor analysis (for reasons explained in the previous chapter), with buffers dimensioned to cope with cell-scale queueing behaviour only. The ATM Forum equivalent is the 'variable bit-rate (VBR) service category'.

Adding a third bandwidth parameter to quantify the burst length allows another form of statistical bit-rate capability. This assumes buffers are large enough to cope with burst-scale queueing, and the CAC algorithm is additionally based on analysis of the burst-scale delay factor. In ATM Forum terminology this is the non-real-time (nrt) VBR service category. However, if the burst length is relatively small, the delays may be small enough to support real-time services.

Note that specifying SBR (VBR) or DBR (CBR) capability does not imply a particular choice of queueing analysis; it just means that the CAC algorithm is required to address both burst-scale and cell-scale queueing components (in the case of SBR/VBR) or just the cell-scale queueing component (in the case of DBR/CBR). Likewise, the bandwidth parameters required in the traffic contract may depend on what analysis is employed (particularly for burst-scale queueing).

ADMISSIBLE LOAD: THE CELL-SCALE CONSTRAINT

Let's say we have dimensioned a buffer to be 40 cells' capacity for a cell loss limit of 10^{-10} and a load of 75% (see Table 10.1). We could make our maximum admissible load 75%, and not accept any more traffic if the extra load would increase the total beyond 75%. But what if the cell loss requirement is not so stringent? In this case the admissible load could be greater than 75%. Some straightforward manipulation of the heavy load

Table 10.1.   CAC Look-up Table for Finite M/D/1: Admissible Load, Given Buffer Capacity and Cell Loss Probability

x (cells)	cell loss probability											
	10^{-1}	10^{-2}	10^{-3}	10^{-4}	10^{-5}	10^{-6}	10^{-7}	10^{-8}	10^{-9}	10^{-10}	10^{-11}	10^{-12}
5	96.3%	59.7%	41.9%	16.6%	6.6%	2.9%	1.35%	0.62%	0.28%	0.13%	0.06%	0.03%
10	99.9%	85.2%	71.2%	60.1%	50.7%	42.7%	35.8%	29.9%	24.9%	20.7%	17.1%	14.2%
15	99.9%	92.4%	82.4%	74.2%	66.9%	60.4%	54.4%	49.0%	44.0%	39.5%	35.4%	31.6%
20	99.9%	95.6%	87.7%	81.3%	75.5%	70.2%	65.2%	60.5%	56.2%	52.1%	48.2%	44.6%
25	99.9%	97.2%	90.7%	85.4%	80.7%	76.2%	72.0%	68.0%	64.2%	60.6%	57.2%	53.9%
30	99.9%	98.2%	92.7%	88.2%	84.1%	80.3%	76.7%	73.2%	69.9%	66.7%	63.6%	60.7%
35	99.9%	98.9%	94.0%	90.1%	86.6%	83.2%	80.0%	77.0%	74.0%	71.2%	68.4%	65.8%
40	99.9%	99.4%	95.0%	91.5%	88.4%	85.4%	82.6%	79.8%	77.2%	74.6%	72.1%	69.7%
45	99.9%	99.7%	95.7%	92.6%	89.8%	87.1%	84.6%	82.1%	79.7%	77.4%	75.1%	72.9%
50	99.9%	99.9%	96.3%	93.5%	90.9%	88.5%	86.2%	83.9%	81.7%	79.6%	77.5%	75.5%
55	99.9%	99.9%	96.7%	94.2%	91.8%	89.6%	87.5%	85.4%	83.4%	81.4%	79.5%	77.6%
60	99.9%	99.9%	97.1%	94.7%	92.6%	90.5%	88.6%	86.7%	84.8%	83.0%	81.2%	79.4%
65	99.9%	99.9%	97.4%	95.2%	93.2%	91.3%	89.5%	87.7%	86.0%	84.3%	82.6%	81.0%
70	99.9%	99.9%	97.7%	95.6%	93.7%	92.0%	90.3%	88.6%	87.0%	85.4%	83.8%	82.3%
75	99.9%	99.9%	97.9%	95.9%	94.2%	92.5%	91.0%	89.4%	87.9%	86.4%	84.9%	83.5%
80	99.9%	99.9%	98.1%	96.2%	94.6%	93.0%	91.5%	90.1%	88.6%	87.2%	85.9%	84.5%
85	99.9%	99.9%	98.2%	96.5%	95.0%	93.5%	92.1%	90.7%	89.3%	88.0%	86.7%	85.4%
90	99.9%	99.9%	98.4%	96.7%	95.3%	93.9%	92.5%	91.2%	89.9%	88.7%	87.4%	86.2%
95	99.9%	99.9%	98.5%	96.9%	95.5%	94.2%	92.9%	91.7%	90.5%	89.3%	88.1%	86.9%
100	99.9%	99.9%	98.6%	97.1%	95.8%	94.5%	93.3%	92.1%	91.0%	89.8%	88.7%	87.6%

approximation for the M/D/1 system (see Chapter 8) gives:

$$\rho = \frac{2 \cdot x}{2 \cdot x - \ln(\text{CLP})}$$

where we have the maximum admissible load defined in terms of the buffer capacity and the cell loss probability requirement.

A CAC algorithm based on M/D/I analysis

How do we use this equation in a CAC algorithm? The traffic contract is based on just two parameters: the peak cell rate, h_i, and the required cell loss probability CLP_i, where $i = 1, 2, \ldots, n$ denotes the set of connections which have already been accepted and are currently in progress, i.e. they have not yet been cleared. Connection $n + 1$ is that request which is currently being tested. This connection is accepted if the following inequality holds:

$$\frac{h_{n+1}}{C} + \sum_{i=1}^{n} \frac{h_i}{C} \leqslant \frac{2 \cdot x}{2 \cdot x - \ln\left(\min_{i=1 \to n+1}(\text{CLP}_i)\right)}$$

where C is the bandwidth capacity of the link. Obviously it is not necessary to perform a summation of the peak rates every time because this can be recorded in a current load variable which is modified whenever a new connection is accepted or an existing connection is cleared. Similarly, a temporary variable holding the most stringent (i.e. the minimum) cell loss probability can be updated whenever a newly accepted connection has a lower CLP. However, care must be taken to ensure that the minimum CLP is recomputed when calls are cleared, so that the performance requirements are based on the current set of accepted connections.

It is important to realize that the cell loss probability is suffered by all admitted connections, because all cells go through the one link in question. Hence the minimum CLP is the one that will give the most stringent limit on the admitted load, and it is this value that is used in the CAC formula. (This is in fact an approximation: different VCs passing through the same 'first-come first-served' link buffer can suffer different cell loss probabilities depending on their particular traffic characteristics, but the variation is not large, and the analysis is complicated.) Priority mechanisms can be used to distinguish between levels of CLP requirements; we deal with this in Chapter 13.

We know that the inequality is based on a heavy traffic approximation. For a buffer size of 40 cells and a CLP requirement of 10^{-10}, the equation gives a maximum admissible load of 77.65%, slightly higher than the 74.6% maximum obtained using the exact analysis. An alternative approach is to use look-up tables based on exact analysis instead of the expression on the right-hand side of the inequality. Table 10.1 shows such a table, giving the maximum percentage load that can be admitted for finite buffer sizes ranging from 5 cells up to 100 cells, and cell loss probabilities ranging from 10^{-1} down to 10^{-10}. This table is generated by iteration of the output buffer analysis of Chapter 7 with Poisson input traffic.

A CAC algorithm based on *N·D/D/1* analysis

But what if all the traffic is CBR and the number of sources is relatively small? We know from the N·D/D/1 analysis that the admissible load can be greater than that given by the M/D/1 results for a given CLP requirement. The problem with the N·D/D/1 analysis is that it models a homogeneous source mix, i.e. all sources have the same traffic characteristics. In general, this will not be the case. However, it turns out that for a fixed load, ρ, and a constant number of sources, N, the worst-case situation for cell loss is the homogeneous case. Thus we can use the N·D/D/1 results and apply them in the general situation where there are N sources of different peak cell rates.

As for the M/D/1 system, we manipulate the heavy load approximation for the N·D/D/1 queue by taking logs of both sides, and rearrange in terms of ρ:

$$\text{CLP} = e^{-2 \cdot x \cdot \left(\frac{x}{N} + \frac{1-\rho}{\rho} \right)}$$

which gives the formula

$$\rho = \frac{2 \cdot x \cdot N}{2 \cdot x \cdot N - (2 \cdot x^2 + N \cdot \ln(\text{CLP}))}$$

It is possible for this formula to return values of admissible load greater than 100%, specifically when

$$2 \cdot x^2 + N \cdot \ln(\text{CLP}) > 0$$

Such a load would obviously take the queue into a permanent (burst-scale) overload, causing significantly more cell loss than that specified. However, it does provide us with a first test for a CAC algorithm based on this analysis, i.e. if

$$n + 1 \leqslant - \frac{2 \cdot x^2}{\ln \left(\min_{i=1 \to n+1} (\text{CLP}_i) \right)}$$

then we can load the link up to 100% with any mix of $n + 1$ CBR sources, i.e. we can accept the connection provided that

$$\frac{h_{n+1}}{C} + \sum_{i=1}^{n} \frac{h_i}{C} \leqslant 1$$

Otherwise, if

$$n + 1 > -\frac{2 \cdot x^2}{\ln\left(\min_{i=1 \to n+1}(\text{CLP}_i)\right)}$$

then we can accept the connection if

$$\frac{h_{n+1}}{C} + \sum_{i=1}^{n} \frac{h_i}{C} \leqslant \frac{2 \cdot x \cdot (n+1)}{2 \cdot x \cdot (n+1) - \left[2 \cdot x^2 + (n+1) \cdot \ln\left(\min_{i=1 \to n+1}(\text{CLP}_i)\right)\right]}$$

It is also important to remember that the N·D/D/1 analysis is only required when $N > x$. If there are fewer sources than buffer places, then the queue never overflows, and so the admissible load is 100%.

Like the M/D/1 system, this inequality is based on a heavy load approximation. A look-up table method based on iteration of the equation

$$\text{CLP} \approx \sum_{n=x+1}^{N} \left\{ \frac{N!}{n! \cdot (N-n)!} \cdot \left(\frac{n-x}{D}\right)^n \cdot \left[1 - \left(\frac{n-x}{D}\right)\right]^{N-n} \cdot \frac{D-N+x}{D-n+x} \right\}$$

provides a *better* approximation than the heavy load approximation, but note that it is not an exact analysis as in Table 10.1 for the finite M/D/1.

The approach is more complicated than for the M/D/1 system because of the dependence on a third parameter, N. Table 10.2 shows the maximum number of sources admissible for a load of 100%, for combinations of buffer capacity and cell loss probability. Table 10.3 then shows the maximum admissible load for combinations of N and cell loss probability, in three parts: (a) for a buffer capacity of 10 cells, (b) for 50 cells, (c) for 100 cells.

The tables are used as follows: first check if the number of sources is less than that given by Table 10.2 for a given CLP and buffer capacity; if so, then the admissible load is 100%. Otherwise, use the appropriate part of Table 10.3, with the given number of sources and CLP requirement, to find the maximum admissible load. Note that when the maximum admissible load is less than 100% of the cell rate capacity of the link, the bandwidth that is effectively being allocated to each source is greater than the source's peak cell rate, h_i. This allocated bandwidth is found simply

Table 10.2. CAC Look-up Table for Deterministic Bit-Rate Transfer Capability: Maximum Number of Sources for 100% Loading, Given Buffer Capacity and Cell Loss Probability

x (cells)	cell loss probability											
	10^{-1}	10^{-2}	10^{-3}	10^{-4}	10^{-5}	10^{-6}	10^{-7}	10^{-8}	10^{-9}	10^{-10}	10^{-11}	10^{-12}
5	23	11	8	6	5	5	5	5	5	5	5	5
10	89	45	30	23	19	16	14	13	12	11	11	10
15	200	100	67	50	41	34	30	26	24	22	20	19
20	353	176	118	89	71	60	52	45	41	37	34	32
25	550	275	183	138	111	92	80	70	63	57	52	48
30	790	395	264	198	159	133	114	100	89	81	74	68
35	1064	537	358	269	215	180	155	136	121	109	100	92
40	1389	701	467	351	281	234	201	176	157	142	129	119
45	1758	886	591	443	355	296	254	223	198	179	163	150
50	2171	1085	729	547	438	365	313	275	244	220	201	185
55	2627	1313	881	661	529	441	379	332	295	266	242	223
60	3126	1563	1042	786	629	525	450	394	351	316	288	264
65	3669	1834	1223	922	738	616	528	462	411	371	337	310
70	4256	2128	1418	1064	856	714	612	536	477	429	391	359
75	4885	2442	1628	1221	982	819	702	615	547	493	448	411
80	5558	2779	1852	1389	1111	931	799	699	622	560	510	468
85	6275	3137	2091	1568	1255	1045	901	789	702	632	575	527
90	7035	3517	2345	1758	1407	1172	1005	884	786	708	644	591
95	7839	3919	2613	1959	1567	1306	1119	985	876	788	717	658
100	8685	4342	2895	2171	1737	1447	1240	1085	970	873	794	729

by dividing the peak cell rate of a source by the maximum admissible load (expressed as a fraction, not as a percentage).

This CAC algorithm, based on either the N·D/D/1 approximate analysis or the associated tables, is appropriate for the deterministic bit-rate capability. The parameters required are just the peak (cell) rate h_i, and the required cell loss probability, CLP_i, for each source i, along with the buffer capacity x, the cell rate capacity C, and the number of connections currently in progress, n. Note that it is acceptable when using the deterministic bit-rate capability to mix variable and constant bit-rate sources, provided that the peak cell rate of a source is used in calculating the allocated load. The important point is that it is only the peak cell rate which is used to characterize the source's traffic behaviour.

The cell-scale constraint in statistical-bit-rate transfer capability, based on M/D/1 analysis

A cell-scale constraint is also a component of the CAC algorithm for the statistical bit-rate transfer capability. Here, the M/D/1 system is more

Table 10.3. (a) Maximum Admissible Load for a Buffer Capacity of 10 Cells, Given Number of Sources and Cell Loss Probability

					cell loss probability							
N	10^{-1}	10^{-2}	10^{-3}	10^{-4}	10^{-5}	10^{-6}	10^{-7}	10^{-8}	10^{-9}	10^{-10}	10^{-11}	10^{-12}
10	100.0%	100.0%	100.0%	100.0%	100.0%	100.0%	100.0%	100.0%	100.0%	100.0%	100.0%	100.0%
11	100.0%	100.0%	100.0%	100.0%	100.0%	100.0%	100.0%	100.0%	100.0%	100.0%	100.0%	84.6%
12	100.0%	100.0%	100.0%	100.0%	100.0%	100.0%	100.0%	100.0%	100.0%	85.7%	70.6%	57.1%
13	100.0%	100.0%	100.0%	100.0%	100.0%	100.0%	100.0%	100.0%	81.3%	68.4%	59.1%	48.2%
14	100.0%	100.0%	100.0%	100.0%	100.0%	100.0%	100.0%	87.5%	73.7%	60.9%	51.9%	42.4%
15	100.0%	100.0%	100.0%	100.0%	100.0%	100.0%	93.8%	79.0%	65.2%	55.6%	46.9%	39.5%
16	100.0%	100.0%	100.0%	100.0%	100.0%	100.0%	84.2%	72.7%	61.5%	51.6%	43.2%	36.4%
17	100.0%	100.0%	100.0%	100.0%	100.0%	94.4%	81.0%	68.0%	56.7%	48.6%	41.5%	34.7%
18	100.0%	100.0%	100.0%	100.0%	100.0%	85.7%	75.0%	64.3%	54.6%	46.2%	39.1%	33.3%
19	100.0%	100.0%	100.0%	100.0%	100.0%	82.6%	73.1%	61.3%	52.8%	44.2%	38.0%	32.2%
20	100.0%	100.0%	100.0%	100.0%	95.2%	80.0%	69.0%	58.8%	50.0%	42.6%	36.4%	30.8%
30	100.0%	100.0%	100.0%	85.7%	75.0%	65.2%	56.6%	48.4%	41.7%	35.7%	30.3%	25.9%
40	100.0%	100.0%	88.9%	78.4%	69.0%	59.7%	51.3%	44.4%	38.1%	32.8%	28.0%	24.0%
50	100.0%	96.2%	84.8%	74.6%	64.9%	56.8%	49.0%	42.4%	36.5%	31.5%	26.9%	22.9%
60	100.0%	93.8%	82.2%	72.3%	63.2%	55.1%	47.6%	41.1%	35.5%	30.5%	26.1%	22.3%
70	100.0%	90.9%	80.5%	70.7%	61.4%	53.9%	46.7%	40.5%	34.8%	29.9%	25.6%	21.9%
80	100.0%	88.9%	78.4%	69.0%	60.6%	53.0%	46.0%	39.8%	34.3%	29.5%	25.3%	21.6%
90	98.9%	88.2%	77.6%	68.2%	60.0%	52.3%	45.5%	39.3%	34.0%	29.2%	25.0%	21.3%
100	98.0%	87.0%	76.9%	67.6%	59.2%	51.8%	44.8%	38.9%	33.7%	28.9%	24.8%	21.2%
200	93.5%	83.0%	73.5%	64.7%	56.7%	49.5%	43.1%	37.4%	32.3%	27.9%	23.9%	20.5%
300	92.0%	81.7%	72.3%	63.7%	56.0%	48.9%	42.6%	37.0%	31.9%	27.5%	23.6%	20.2%
400	91.3%	81.1%	71.8%	63.3%	55.6%	48.5%	42.3%	36.7%	31.7%	27.3%	23.5%	20.1%
500	90.9%	80.8%	71.6%	63.1%	55.3%	48.4%	42.1%	36.6%	31.6%	27.3%	23.4%	20.0%
600	90.6%	80.5%	71.5%	62.8%	55.2%	48.2%	42.0%	36.5%	31.6%	27.2%	23.3%	20.0%
700	90.4%	80.4%	71.4%	62.7%	55.1%	48.1%	41.9%	36.4%	31.5%	27.1%	23.3%	20.0%
800	90.3%	80.2%	71.2%	62.6%	55.0%	48.1%	41.9%	36.4%	31.5%	27.1%	23.3%	19.9%
900	90.2%	80.1%	71.1%	62.5%	54.9%	48.0%	41.8%	36.3%	31.4%	27.1%	23.3%	19.9%
1000	90.1%	80.1%	71.0%	62.5%	54.9%	48.0%	41.8%	36.3%	31.4%	27.1%	23.2%	19.9%

appropriate, using the mean cell rate, m_i, instead of the peak cell rate h_i, to calculate the load in the inequality test; i.e. if

$$\frac{m_{n+1}}{C} + \sum_{i=1}^{n} \frac{m_i}{C} \leqslant \frac{2 \cdot x}{2 \cdot x - \ln\left(\min_{i=1 \to n+1}(CLP_i)\right)}$$

is satisfied, then the cell-scale behaviour is within the required cell loss probability limits, and the CAC algorithm must then check the burst-scale constraint before making an accept/reject decision. If the inequality is not satisfied, then the connection can immediately be rejected. For a more accurate test, values from the look-up table in Table 10.1 can be used instead of the expression on the right-hand side of the inequality.

Table 10.3. (b) Maximum Admissible Load for a Buffer Capacity of 50 Cells, Given Number of Sources and Cell Loss Probability

					cell loss probability							
N	10^{-1}	10^{-2}	10^{-3}	10^{-4}	10^{-5}	10^{-6}	10^{-7}	10^{-8}	10^{-9}	10^{-10}	10^{-11}	10^{-12}
180	100.0%	100.0%	100.0%	100.0%	100.0%	100.0%	100.0%	100.0%	100.0%	100.0%	100.0%	100.0%
190	100.0%	100.0%	100.0%	100.0%	100.0%	100.0%	100.0%	100.0%	100.0%	100.0%	100.0%	99.0%
200	100.0%	100.0%	100.0%	100.0%	100.0%	100.0%	100.0%	100.0%	100.0%	100.0%	100.0%	97.1%
210	100.0%	100.0%	100.0%	100.0%	100.0%	100.0%	100.0%	100.0%	100.0%	100.0%	98.6%	95.9%
220	100.0%	100.0%	100.0%	100.0%	100.0%	100.0%	100.0%	100.0%	100.0%	100.0%	97.4%	94.8%
240	100.0%	100.0%	100.0%	100.0%	100.0%	100.0%	100.0%	100.0%	100.0%	97.6%	95.2%	92.7%
260	100.0%	100.0%	100.0%	100.0%	100.0%	100.0%	100.0%	100.0%	98.5%	95.9%	93.5%	90.9%
280	100.0%	100.0%	100.0%	100.0%	100.0%	100.0%	100.0%	99.3%	96.9%	94.6%	92.1%	89.7%
300	100.0%	100.0%	100.0%	100.0%	100.0%	100.0%	100.0%	98.0%	95.5%	93.2%	90.9%	88.5%
350	100.0%	100.0%	100.0%	100.0%	100.0%	100.0%	98.0%	95.6%	93.1%	90.9%	88.6%	86.2%
400	100.0%	100.0%	100.0%	100.0%	100.0%	98.5%	96.2%	93.9%	91.5%	89.1%	87.0%	84.8%
450	100.0%	100.0%	100.0%	100.0%	99.6%	97.2%	94.7%	92.4%	90.2%	87.9%	85.7%	83.5%
500	100.0%	100.0%	100.0%	100.0%	98.4%	96.0%	93.6%	91.4%	89.1%	87.0%	84.8%	82.5%
550	100.0%	100.0%	100.0%	99.8%	97.5%	95.2%	92.8%	90.5%	88.3%	86.1%	84.0%	81.9%
600	100.0%	100.0%	100.0%	99.0%	96.6%	94.3%	92.0%	89.8%	87.6%	85.5%	83.3%	81.2%
700	100.0%	100.0%	100.0%	97.9%	95.5%	93.2%	90.9%	88.7%	86.5%	84.4%	82.4%	80.3%
800	100.0%	100.0%	99.3%	97.0%	94.7%	92.4%	90.2%	87.9%	85.8%	83.7%	81.6%	79.6%
900	100.0%	100.0%	98.6%	96.3%	94.0%	91.7%	89.6%	87.4%	85.2%	83.1%	81.1%	79.0%
1000	100.0%	100.0%	98.1%	95.7%	93.5%	91.2%	89.1%	86.9%	84.8%	82.6%	80.7%	78.6%

Table 10.3. (c) Maximum Admissible Load for a Buffer Capacity of 100 Cells

	cell loss probability				
N	10^{-8}	10^{-9}	10^{-10}	10^{-11}	10^{-12}
700	100.0%	100.0%	100.0%	100.0%	100.0%
750	100.0%	100.0%	100.0%	100.0%	99.5%
800	100.0%	100.0%	100.0%	99.9%	98.6%
850	100.0%	100.0%	100.0%	99.1%	97.8%
900	100.0%	100.0%	99.6%	98.4%	97.2%
950	100.0%	100.0%	99.0%	97.7%	96.5%
1000	100.0%	99.6%	98.4%	97.2%	96.0%

ADMISSIBLE LOAD: THE BURST SCALE

Let's now look at the loads that can be accepted for bursty sources. For this we will use the burst-scale loss analysis of the previous chapter, i.e. assume that the buffer is of zero size at the burst scale. Remember that each source has an average rate of m cell/s; so, with N sources, the utilization is given by

$$\rho = \frac{N \cdot m}{C}$$

Unfortunately we do not have a simple approximate formula that can be manipulated to give the admissible load as an explicit function of the traffic contract parameters. The best we can do to simplify the situation is to use the approximate formula for the burst-scale loss factor:

$$\text{CLP} \approx \frac{1}{(1-\rho)^2 \cdot N_0} \cdot \frac{(\rho \cdot N_0)^{\lfloor N_0 \rfloor}}{\lfloor N_0 \rfloor!} \cdot e^{-\rho \cdot N_0}$$

How can we use this formula in a connection admission control algorithm? In a similar manner to Erlang's lost call formula, we must use the formula to produce a table which allows us, in this case, to specify the required cell loss probability and the source peak cell rate and find out the maximum allowed utilization. We can then calculate the maximum number of sources of this type (with mean cell rate m) that can be accepted using the formula

$$N = \frac{\rho \cdot C}{m}$$

Table 10.4 does not directly use the peak cell rate, but, rather, the number of peak cell rates which fit into the service capacity, i.e. the parameter N_0. Example peak rates for the standard service capacity of 353 208 cell/s are shown.

So, if we have a source with a peak cell rate of 8830.19 cell/s (i.e. 3.39 Mbit/s) and a mean cell rate of 2000 cell/s (i.e. 768 kbit/s), and we want the CLP to be no more than 10^{-10}, then we can accept

Table 10.4. Maximum Admissible Load for Burst-Scale Constraint

h (cell/s)	N_0	cell loss probability											
		10^{-1}	10^{-2}	10^{-3}	10^{-4}	10^{-5}	10^{-6}	10^{-7}	10^{-8}	10^{-9}	10^{-10}	10^{-11}	10^{-12}
35 320.76	10	72.1%	52.3%	37.9%	28.1%	21.2%	16.2%	12.5%	9.7%	7.6%	5.9%	4.7%	3.7%
17 660.38	20	82.3%	67.0%	54.3%	44.9%	37.7%	32.0%	27.4%	23.6%	20.5%	17.8%	15.6%	13.6%
11 773.59	30	86.5%	73.7%	62.5%	53.8%	46.9%	41.4%	36.8%	32.9%	29.6%	26.7%	24.1%	21.9%
8 830.19	40	88.9%	77.8%	67.5%	59.5%	53.0%	47.7%	43.3%	39.4%	36.1%	33.2%	30.6%	28.2%
7 064.15	50	90.5%	80.5%	71.1%	63.5%	57.4%	52.4%	48.1%	44.3%	41.1%	38.2%	35.6%	33.2%
5 886.79	60	91.7%	82.5%	73.7%	66.6%	60.8%	55.9%	51.8%	48.2%	45.0%	42.2%	39.6%	37.3%
5 045.82	70	92.5%	84.1%	75.8%	69.0%	63.5%	58.8%	54.8%	51.3%	48.3%	45.5%	43.0%	40.7%
4 415.09	80	93.2%	85.3%	77.4%	71.0%	65.7%	61.2%	57.3%	54.0%	51.0%	48.3%	45.8%	43.6%
3 924.53	90	93.7%	86.3%	78.8%	72.6%	67.5%	63.2%	59.5%	56.2%	53.3%	50.6%	48.2%	46.0%
3 532.08	100	94.2%	87.2%	80.0%	74.0%	69.1%	64.9%	61.3%	58.1%	55.3%	52.7%	50.4%	48.2%
1 766.04	200	96.4%	91.7%	86.4%	81.8%	78.0%	74.7%	71.8%	69.3%	67.0%	64.9%	62.9%	61.1%
1 177.36	300	97.3%	93.6%	89.2%	85.3%	82.0%	79.2%	76.8%	74.6%	72.6%	70.7%	69.0%	67.5%
883.02	400	97.8%	94.7%	90.8%	87.4%	84.5%	82.0%	79.8%	77.8%	76.0%	74.4%	72.8%	71.4%
706.42	500	98.1%	95.4%	91.9%	88.8%	86.2%	83.9%	81.9%	80.1%	78.4%	76.9%	75.5%	74.2%
588.68	600	98.4%	95.9%	92.7%	89.9%	87.4%	85.3%	83.4%	81.8%	80.2%	78.8%	77.5%	76.3%
504.58	700	98.5%	96.3%	93.3%	90.7%	88.4%	86.4%	84.7%	83.1%	81.7%	80.3%	79.1%	78.0%

$$N = \frac{0.332 \times 353\,208}{2000} = 58.63$$

i.e. 58 connections of this type. This is 18 more connections than if we had used the deterministic bit-rate capability (assuming 100% allocation of peak rates, which is possible if the buffer capacity is 25 cells or more).

The ratio

$$G = \frac{N}{N_0}$$

is called the 'statistical multiplexing gain'. This is the actual number accepted, N, divided by the number N_0 if we were to allocate on the peak rate only. It gives an indication of how much better the utilization is when using SBR capability compared with using DBR capability. If peak rate allocation is used, then there is no statistical multiplexing gain, and G is 1.

But what happens if there are different types of source? If all the sources have the same peak cell rate, then the actual mean rates of individual sources do not matter, so long as the total mean cell rate is less than $\rho \cdot C$, i.e.

$$\sum_i m_i \leqslant \rho \cdot C$$

So, the connection is accepted if the following inequality holds:

$$\frac{m_{n+1}}{C} + \sum_{i=1}^{n} \frac{m_i}{C} \leqslant \rho(\text{CLP}, N_0)$$

where ρ is chosen (as a function of CLP and N_0) from Table 10.4 in the manner described previously.

A practical CAC scheme

Notice in the table that the value of ρ decreases as the peak cell rate increases. We could therefore use this approach in a more conservative way by choosing ρ according to the most stringent (i.e. highest) peak-rate source in the mix. This is effectively assuming that all sources, whatever their mean rate, have a peak rate equal to the highest in the traffic mix. The CAC algorithm would need to keep track of this maximum peak rate (as well as the minimum CLP requirement), and update the admissible load accordingly. The inequality test for this scheme is therefore written as:

$$\frac{m_{n+1}}{C} + \sum_{i=1}^{n} \frac{m_i}{C} \leqslant \rho \left(\min_{i=1 \rightarrow n+1} (\text{CLP}_i), \frac{C}{\max_{i=1 \rightarrow n+1} (h_i)} \right)$$

Equivalent cell rate and linear CAC

A different approach is to think in terms of the cell rate allocated to a source. For the DBR capability, a CAC algorithm allocates either the source's peak cell rate or a value *greater* than this, because cell-scale queueing limits the admissible load. This keeps the utilization, defined in terms of peak rates, at or below 100%. With SBR capability, the total *peak* rate allocated can be in excess of 100%, so the actual portion of service capacity allocated to a source is below the peak cell rate (and, necessarily, above the mean cell rate). This allocation is called the *equivalent cell rate*.

Other terms have been used to describe essentially the same concept: 'effective bandwidth' and 'equivalent capacity' are the most common terms used, but the precise definition is usually associated with a particular analytical method. 'Equivalent cell rate' is the term used in the ITU Standards documents.

The key contribution made by the concept of equivalent cell rate is the idea of a single value to represent the amount of resource required for a single source in a traffic mix at a given CLP requirement. This makes the admission control process simply a matter of adding the equivalent cell rate of the requested connection to the currently allocated value. If it exceeds the service rate available then the request is rejected. This is an attractive approach for traffic mixes of different types of sources because of its apparent simplicity. It is known as 'linear CAC'.

The difficulty lies in defining the equivalent cell rate for a particular source type. The issue rests on how well different types of sources are able to mix when multiplexed through the same buffer. The analysis we have used so far is for a traffic mix of sources of the same type. In this case, the equivalent cell rate can be defined as

$$\text{ECR} = \frac{C}{N} = \frac{C}{h} \cdot \frac{h}{N} = \frac{N_0}{N} \cdot h = \frac{h}{G}$$

When the statistical multiplexing gain, G, is low (i.e. approaching a value of 1), the equivalent cell rate approaches the peak rate of the source and the cell loss probability will be low. Conversely, when the gain is high, the equivalent cell rate approaches the mean rate of the source, and the cell loss probability is high.

Equivalent cell rate based on a traffic mix of sources of the same type may underestimate the resources required when sources of very different characteristics are present. The exact analysis of heterogeneous source multiplexing is beyond the scope of this book, but there are other approaches.

Two-level CAC

One of these approaches, aimed at simplifying CAC, is to divide the sources into classes and partition the service capacity so that each source class is allocated a proportion of it. The homogeneous source analysis can be justified in this case because the fraction of service rate allocated to the *class* is used instead of the total service capacity (within the fraction, all the sources being the same). This has the effect of reducing N_0, and hence reducing the admissible load per class. The problem with this approach is that a connection of one type may be rejected if its allocation is full even though there is unused capacity because other service classes are underused.

A solution is to divide the CAC algorithm into two levels. The first level makes accept/reject decisions by comparing the current service-class allocations with the maximum number allowed. But this is supported by a second-level 'back-room' task which redistributes unused capacity to service classes that need it. The second level is computationally intensive because it must ensure that the allocations it proposes conform to the required cell loss probability. This takes time, and so the allocations are updated on a (relatively) longer time scale. However, the first level is a very simple comparison and so a connection request can be assessed immediately.

The basic principle of the two-level scheme is to have a first level which can make an instant decision on a connection request, and a second level which can perform detailed traffic calculations in the background to keep the scheme as accurate as possible. The service class approach is just one of many: other algorithms for the first and second levels have been proposed in the literature.

Accounting for the burst-scale delay factor

Whatever the size of buffer, large or small, the actual burst-scale loss depends on the two burst-scale factors: the loss factor assumes there is no buffer, and the delay factor quantifies how much less is the loss if we incorporate buffering. Thus if we use the loss factor only, we will tend to overestimate the cell loss; or for a fixed CLP, we will underestimate the admissible load.

So, for small buffer capacities, just using the loss factor is a good starting point for admission control at the burst scale. But we have already incorporated some 'conservative' assumptions into our practical scheme, and even small buffers can produce some useful gains under certain circumstances. How can the scheme be modified to account for the burst-scale delay factor, and hence increase the admissible load?

Let's use our previous example of 58 connections (peak cell rate 8830.19 cell/s, mean cell rate 2000 cell/s, $N_0 = 40$, and a CLP of 10^{-10}) and see how many more connections can be accepted if the average burst duration is 40 ms and the buffer capacity is 475 cells. First, we need to calculate:

$$\frac{N_0 \cdot x}{b} = \frac{40 \times 475}{0.04 \times 8830.19} = 53.79$$

and the admissible load (from the burst-scale loss analysis) is

$$\rho = \frac{58 \times 2000}{353\,208} = 0.328$$

So we can calculate the CLP gain due to the burst-scale delay factor:

$$\mathrm{CLP}_{\mathrm{excess\text{-}rate}} = e^{-\left[N_0 \cdot \frac{X}{b} \cdot \frac{(1-\rho)^3}{4 \cdot \rho + 1}\right]} = 8.58 \times 10^{-4}$$

Thus there is a further CLP gain of about 10^{-3}, i.e. an overall CLP of about 10^{-13}.

Although the excess-rate cell loss is an exponential function, which can thus be rearranged fairly easily, we will use a tabular approach because it clearly illustrates the process required. Table 10.5 specifies the CLP and the admissible load in order to find a value for $N_0 \cdot x/b$ (this was introduced in Chapter 9 as the size of a buffer in units of excess-rate bursts). The CLP target is 10^{-10}. By how much can the load be increased so that the overall CLP meets this target? Looking down the 10^{-2} column of Table 10.5, we find that the admissible load could increase to a value of nearly 0.4. Then, we check in Table 10.4 to see that the burst-scale loss contribution for a load of 0.394 is 10^{-8}. Thus the overall CLP meets our target of 10^{-10}.

The number of connections that can be accepted is now

$$N = \frac{0.394 \times 353\,208}{2000} = 69.58$$

i.e. 69 connections of this type. This is a further 11 connections more than if we had just used the burst-scale loss factor as the basis for the CAC algorithm. The penalty is the increased size of the buffer, and the correspondingly greater delays incurred (about 1.3 ms maximum, for a buffer capacity of 475 cells). However, the example illustrates the principle, and even with buffers of less than 100 cells, worthwhile gains in admissible load are possible. The main difficulty with the process is in selecting a load to provide cell loss factors from Tables 10.4 and 10.5 which combine to the required target cell loss. The target cell loss can be found by trial and error, gradually reducing the excess rate CLP by taking the next column to the left in Table 10.5.

Table 10.5. Burst-Scale Delay Factor Table for Values of $N_0 \cdot x/b$, Given Admissible Load and CLP

load, ρ	cell loss probability											
	10^{-1}	10^{-2}	10^{-3}	10^{-4}	10^{-5}	10^{-6}	10^{-7}	10^{-8}	10^{-9}	10^{-10}	10^{-11}	10^{-12}
0.02	2.6	5.3	7.9	10.6	13.2	15.9	18.5	21.1	23.8	26.4	29.1	31.7
0.04	3.0	6.0	9.1	12.1	15.1	18.1	21.1	24.2	27.2	30.2	33.2	36.2
0.06	3.4	6.9	10.3	13.8	17.2	20.6	24.1	27.5	30.9	34.4	37.8	41.3
0.08	3.9	7.8	11.7	15.6	19.5	23.4	27.3	31.2	35.1	39.0	42.9	46.8
0.10	4.4	8.8	13.3	17.7	22.1	26.5	31.0	35.4	39.8	44.2	48.6	53.1
0.12	5.0	10.0	15.0	20.0	25.0	30.0	35.0	40.0	45.0	50.0	55.0	60.0
0.14	5.6	11.3	16.9	22.6	28.2	33.9	39.5	45.2	50.8	56.5	62.1	67.8
0.16	6.4	12.7	19.1	25.5	31.9	38.2	44.6	51.0	57.3	63.7	70.1	76.5
0.18	7.2	14.4	21.5	28.7	35.9	43.1	50.3	57.5	64.6	71.8	79.0	86.2
0.20	8.1	16.2	24.3	32.4	40.5	48.6	56.7	64.8	72.9	81.0	89.0	97.1
0.22	9.1	18.2	27.4	36.5	45.6	54.7	63.9	73.0	82.1	91.2	100.3	109.5
0.24	10.3	20.6	30.8	41.1	51.4	61.7	72.0	82.2	92.5	102.8	113.1	123.4
0.26	11.6	23.2	34.8	46.4	58.0	69.6	81.1	92.7	104.3	115.9	127.5	139.1
0.28	13.1	26.2	39.2	52.3	65.4	78.5	91.5	104.6	117.7	130.8	143.9	156.9
0.30	14.8	29.5	44.3	59.1	73.8	88.6	103.4	118.2	132.9	147.7	162.5	177.2
0.32	16.7	33.4	50.1	66.8	83.5	100.2	116.9	133.6	150.3	167.0	183.7	200.4
0.34	18.9	37.8	56.7	75.6	94.5	113.4	132.3	151.2	170.1	189.0	207.9	226.8
0.36	21.4	42.9	64.3	85.7	107.2	128.6	150.0	171.5	192.9	214.3	235.8	257.2
0.38	24.3	48.7	73.0	97.4	121.7	146.1	170.4	194.8	219.1	243.5	267.8	292.2
0.40	27.7	55.4	83.1	110.9	138.6	166.3	194.0	221.7	249.4	277.2	304.9	332.6
0.42	31.6	63.3	94.9	126.5	158.1	189.8	221.4	253.0	284.6	316.3	347.9	379.5
0.44	36.2	72.4	108.6	144.8	180.9	217.1	253.3	289.5	325.7	361.9	398.1	434.3
0.46	41.5	83.1	124.6	166.1	207.6	249.2	290.7	332.2	373.8	415.3	456.8	498.3
0.48	47.8	95.6	143.5	191.3	239.1	286.9	334.7	382.5	430.4	478.2	526.0	573.8
0.50	55.3	110.5	165.8	221.0	276.3	331.6	386.8	442.1	497.4	552.6	607.9	663.1
0.52	64.1	128.3	192.4	256.5	320.6	384.8	448.9	513.0	577.1	641.3	705.4	769.5
0.54	74.8	149.5	224.3	299.0	373.8	448.5	523.3	598.0	672.8	747.5	822.3	897.0

(continued overleaf)

Table 10.5. (continued)

load, ρ	cell loss probability											
	10^{-1}	10^{-2}	10^{-3}	10^{-4}	10^{-5}	10^{-6}	10^{-7}	10^{-8}	10^{-9}	10^{-10}	10^{-11}	10^{-12}
0.56	87.6	175.2	262.7	350.3	437.9	525.5	613.1	700.6	788.2	875.8	963.4	1051.0
0.58	103.2	206.4	309.5	412.7	515.9	619.1	722.3	825.5	928.6	1031.8	1135.0	1238.2
0.60	122.3	244.6	367.0	489.3	611.6	733.9	856.3	978.6	1100.9	1223.2	1345.6	1467.9
0.62	146.0	292.1	438.1	584.1	730.2	876.2	1022.2	1168.2	1314.3	1460.3	1606.3	1752.4
0.64	175.7	351.4	527.1	702.8	878.5	1054.2	1229.9	1405.6	1581.3	1756.9	1932.6	2108.3
0.66	213.2	426.5	639.7	853.0	1066.2	1279.5	1492.7	1706.0	1919.2	2132.5	2345.7	2558.9
0.68	261.4	522.8	784.2	1045.6	1307.0	1568.4	1829.8	2091.2	2352.6	2614.0	2875.4	3136.8
0.70	324.1	648.1	972.2	1296.3	1620.3	1944.4	2268.5	2592.5	2916.6	3240.7	3564.7	3888.8
0.72	407.0	814.0	1220.9	1627.9	2034.9	2441.9	2848.9	3255.8	3662.8	4069.8	4476.8	4883.8
0.74	518.8	1037.6	1556.4	2075.2	2593.9	3112.7	3631.5	4150.3	4669.1	5187.9	5706.7	6225.5
0.76	672.9	1345.8	2018.8	2691.7	3364.6	4037.5	4710.4	5383.4	6056.3	6729.2	7402.1	8075.0
0.78	890.9	1781.9	2672.8	3563.7	4454.7	5345.6	6236.5	7127.5	8018.4	8909.3	9800.3	10691.2
0.80	1208.9	2417.7	3626.6	4835.4	6044.3	7253.1	8462.0	9670.9	10879.7	12088.6	13297.4	14506.3
0.82	1689.8	3379.7	5069.5	6759.3	8449.1	10139.0	11828.8	13518.6	15208.4	16898.3	18588.1	20277.9
0.84	2451.0	4902.0	7353.0	9804.0	12255.0	14706.0	17157.0	19608.0	22058.9	24509.9	26960.9	29411.9
0.86	3725.8	7451.5	11177.3	14903.0	18628.8	22354.5	26080.3	29806.1	33531.8	37257.6	40983.3	44709.1
0.88	6023.0	12045.9	18068.9	24091.9	30114.8	36137.8	42160.8	48183.7	54206.7	60229.7	66252.6	72275.6
0.90	10591.9	21183.8	31775.7	42367.6	52959.5	63551.3	74143.2	84735.1	95327.0	105918.9	116510.8	127102.7
0.92	21047.1	42094.1	63141.2	84188.3	105235.3	126282.4	147329.5	168376.5	189423.6	210470.7	231517.7	252564.8
0.94	50742.2	101484.3	152226.5	202968.6	253710.8	304452.9	355195.1	405937.2	456679.4	507421.5	558163.7	608905.8
0.96	174133.0	348266.0	522399.0	696532.0	870665.0	1044798.0	1218931.0	1393064.0	1567197.0	1741330.0	1915463.0	2089596.0
0.98	1416089.8	2832179.7	4248269.5	5664359.3	7080449.2	8496539.0	9912628.8	11328718.7	12744808.5	14160898.3	15576988.2	16993078.0

CAC IN THE STANDARDS

Connection admission control is defined in ITU Recommendation I.371 [10.1] as the set of actions taken by the network at the call set-up phase (or during call re-negotiation) to establish whether a connection can be accepted or whether it must be rejected. The wording in the ATM Forum Traffic Management Specification 4.1 [10.2] is very similar. We have seen that the CAC algorithm needs to know the source traffic characteristics and the required performance in order to determine whether the connection can be accepted or not and, if accepted, the amount of network resources to allocate. Also it must set the traffic parameters needed by usage parameter control – this will be addressed in the next chapter.

Neither Recommendation I.371 nor the Traffic Management Specification specifies any particular CAC algorithm; they merely observe that many CAC policies are possible, and it is up to the network operator to choose. ITU Recommendation E.736 outlines some possible policies [10.3]. It distinguishes three different operating principles:

1. multiplexing of constant-bit-rate streams
2. rate-envelope multiplexing
3. rate-sharing statistical multiplexing

The first corresponds to peak rate allocation, i.e. the deterministic bit-rate transfer capability, and deals with the cell-scale queueing behaviour. In this book we have considered two different algorithms, based on either the $M/D/1$ or $N \cdot D/D/1$ systems. The second and third operating principles allow for the statistical multiplexing of variable bit-rate streams and are two approaches to providing the statistical bit-rate transfer capability. 'Rate envelope multiplexing' is the term for what we have called the 'burst-scale loss factor', i.e. it is the bufferless approach. The term arises because the objective is to keep the total input rate to within the service rate; any excess rate is assumed to be lost. Rate sharing corresponds to the combined burst-scale loss and delay factors, i.e. it assumes there is a large buffer available to cope with the excess cell rates. It allows higher admissible loads, but the penalty is greater delay. Thus the objective is not to limit the combined cell rate, but to share the service capacity by providing sufficient buffer space to absorb the excess-rate cells.

These three different operating principles require different traffic parameters to describe the source traffic characteristics. DBR requires just the peak cell rate of the source. Rate envelope multiplexing additionally needs the mean cell rate, and rate sharing requires peak cell rate, mean cell rate and some measure of burst length. The actual parameters depend on the CAC policy and what information it uses. But there is one important principle that applies regardless of the policy: if a CAC policy depends on a particular traffic parameter, then the network operator needs to ensure that the value the user has declared for that parameter is not exceeded during the actual flow of cells from the source. Only then can the network operator be confident that the performance requirements will be guaranteed. This is the job of usage parameter control.

11 **Usage Parameter Control**

there's a hole in my bucket

PROTECTING THE NETWORK

We have discussed the statistical multiplexing of traffic through ATM buffers and connection admission control mechanisms to limit the number of simultaneous connections, but how do we know that a traffic source is going to conform to the parameter values used in the admission control decision? There is nothing to stop a source sending cells over the access link at a far higher rate. It is the job of usage parameter control to ensure that any cells over and above the agreed values do not get any further into the network. These agreed values, including the performance requirements, are called the 'traffic contract'.

Usage parameter control is defined as the set of actions taken by the network, at the user access, to monitor and control traffic in terms of conformity with the agreed traffic contract. The main purpose is to protect network resources from source traffic misbehaviour that could affect the quality of service of other established connections. UPC does this by detecting violations of negotiated parameters and taking appropriate actions, for example discarding or tagging cells, or clearing the connection.

A specific control algorithm has not been standardized – as with CAC algorithms, the network may use any algorithm for UPC. However, any such control algorithm should have the following desirable features:

- the ability to detect any traffic situation that does not conform to the traffic contract,
- a rapid response to violations of the traffic contract, and
- being simple to implement.

But are all these features possible in one algorithm? Let's recall what parameters we want to check. The most important one is the peak cell rate; it is needed for both deterministic and statistical bit-rate transfer capabilities. For SBR, the traffic contract also contains the mean cell rate (for rate envelope multiplexing). With rate-sharing statistical multiplexing, the burst length is additionally required. Before we look at a specific algorithm, let's consider the feasibility of controlling the mean cell rate.

CONTROLLING THE MEAN CELL RATE

Suppose we count the total number of cells being sent in some 'measurement interval', T, by a Poisson source. The source has a declared mean cell rate, λ, of one cell per time unit. Is it correct to allow no more than one cell per time unit into the network? We know from Chapter 6 that the probability of k cells arriving in one time unit from a Poisson source is given by

$$\Pr\{k \text{ arrivals in one time unit}\} = \frac{(\lambda \cdot T)^k}{k!} \cdot e^{-\lambda \cdot T}$$

So the probability of more than one arrival per time unit is

$$= 1 - \frac{(1)^0}{0!} \cdot e^{-1} - \frac{(1)^1}{1!} \cdot e^{-1} = 0.2642$$

Thus this strict mean cell rate control would reject one or more cells from a well-behaved Poisson source in 26 out of every 100 time units. What proportion of the number of cells does this represent? Well, we know that the mean number of cells per time unit is 1, and this can also be found by summing the probabilities of there being k cells weighted by the number of cells, k, i.e.

$$\text{mean number of cells} = 1 = 0 \cdot \frac{(1)^0}{0!} \cdot e^{-1} + 1 \cdot \frac{(1)^1}{1!} \cdot e^{-1} + 2 \cdot \frac{(1)^2}{2!} \cdot e^{-1}$$

$$+ \cdots + k \cdot \frac{(1)^k}{k!} \cdot e^{-1} + \cdots$$

When there are $k \geqslant 1$ cell arrivals in a time unit, then one cell is allowed on to the network and $k - 1$ are rejected. Thus the proportion of cells being allowed on to the network is

$$\frac{1 \cdot \frac{(1)^1}{1!} \cdot e^{-1} + \sum_{k=2}^{\infty} 1 \cdot \frac{(1)^k}{k!} \cdot e^{-1}}{1} = 0.6321$$

which means that almost 37% of cells are being rejected although the traffic contract is not being violated.

There are two options open to us: increase the maximum number of cells allowed into the network per time unit or increase the measurement interval to many time units. The object is to decrease this proportion of cells being rejected to an acceptably low level, for example 1 in 10^{10}.

Let's define j as the maximum number of cells allowed into the network during time interval T. The first option requires us to find the smallest value of j for which the following inequality holds:

$$\frac{\sum_{k=j+1}^{\infty}\left\{(k-j)\cdot\frac{(\lambda\cdot T)^k}{k!}\cdot e^{-\lambda\cdot T}\right\}}{\lambda\cdot T}\leqslant 10^{-10}$$

where, in this case, the mean cell rate of the source, λ, is 1 cell per time unit, and the measurement interval, T, is 1 time unit. Table 11.1 shows the proportion of cells rejected for a range of values of j.

To meet our requirement of no more than 1 in 10^{10} cells rejected for a Poisson source of mean rate 1 cell per time unit, we must accept up to 12 cells per time unit. If the Poisson source doubles its rate, then our limit of 12 cells per time unit would result in 1.2×10^{-7} of the cells being rejected. Ideally we would want 50% of the cells to be rejected to keep the source to its contracted mean of 1 cell per time unit. If the Poisson source increases its rate to 10 cells per time unit, then 5.3% of the cells are

Table 11.1. Proportion of Cells Rejected when no more than j cells Are Allowed per Time Unit

j	proportion of cells rejected for a mean cell rate of		
	1 cell/time unit	2 cells/time unit	10 cells/time unit
1	3.68E-01	5.68E-01	9.00E-01
2	1.04E-01	2.71E-01	8.00E-01
3	2.33E-02	1.09E-01	7.00E-01
4	4.35E-03	3.76E-02	6.01E-01
5	6.89E-04	1.12E-02	5.04E-01
6	9.47E-05	2.96E-03	4.11E-01
7	1.15E-05	6.95E-04	3.24E-01
8	1.25E-06	1.47E-04	2.46E-01
9	1.22E-07	2.82E-05	1.79E-01
10	1.09E-08	4.96E-06	1.25E-01
11	9.00E-10	8.03E-07	8.34E-02
12	6.84E-11	1.21E-07	5.31E-02
13	4.84E-12	1.69E-08	3.22E-02
14	3.20E-13	2.21E-09	1.87E-02
15	1.98E-14	2.71E-10	1.03E-02

rejected, and hence over 9 cells per time unit are allowed through. Thus measurement over a short interval means that either too many legitimate cells are rejected (if the limit is small) or, for cells which violate the contract, not enough are rejected (when the limit is large).

Let's now extend the measurement interval. Instead of tabulating for all values of j, the results are shown in Figure 11.1 for two different time intervals: 10 time units and 100 time units. For the 10^{-10} requirement, j is 34 (for $T = 10$) and 163 (for $T = 100$), i.e. the rate is limited to 3.4 cells per time unit, or 1.63 cells per time unit over the respective measurement intervals. So, as the measurement interval increases, the mean rate is being more closely controlled. The problem now is that the time taken to

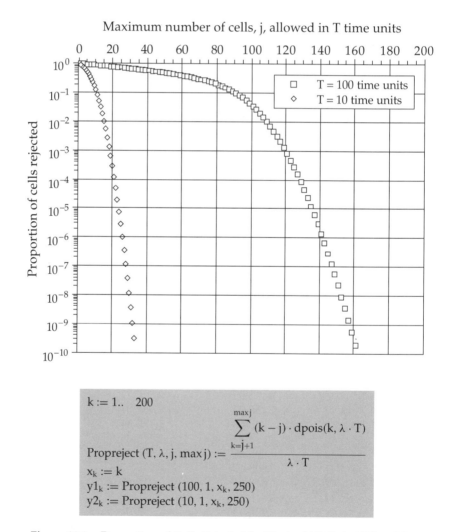

$$k := 1..\ 200$$

$$\text{Propreject}\ (T, \lambda, j, \max j) := \frac{\displaystyle\sum_{k=j+1}^{\max j} (k-j) \cdot \text{dpois}(k, \lambda \cdot T)}{\lambda \cdot T}$$

$$x_k := k$$
$$y1_k := \text{Propreject}\ (100, 1, x_k, 250)$$
$$y2_k := \text{Propreject}\ (10, 1, x_k, 250)$$

Figure 11.1. Proportion of Cells Rejected for Limit of j Cells in T Time Units

respond to violations of the contract is longer. This can result in action being taken too late to protect the network from the effect of the contract violation.

Figure 11.2 shows how the limit on the number of cells allowed per time unit varies with the measurement interval, for a rejection probability of 10^{-10}. The shorter the interval, the poorer the control of the mean rate because of the large 'safety margin' required. The longer the interval, the slower the response to violations of the contract.

So we see that mean cell rate control requires a safety margin between the controlled cell rate and the negotiated cell rate to cope with the

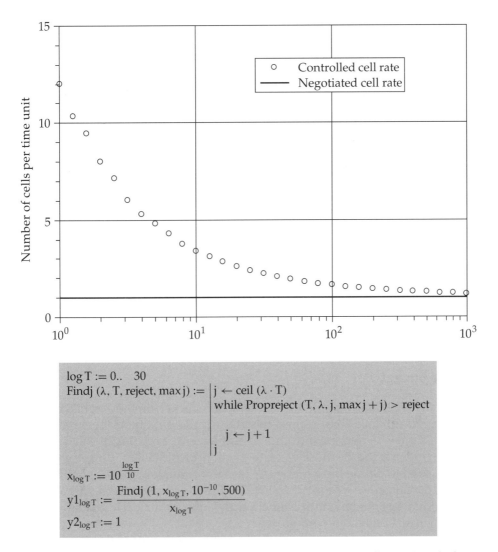

Figure 11.2. Controlling the Mean Cell Rate over Different Time Scales

statistical fluctuations of well-behaved traffic streams, but this safety margin limits the ability of the UPC function to detect violations of the negotiated mean cell rate. As the measurement interval is extended, the safety margin required becomes less, but then any action in response to contract violation may be too late to be an effective protection for network resources.

Therefore we need to modify how we think of the mean cell rate: it is necessary to think in terms of a 'virtual mean' defined over some specified time interval. The compromise is between the accuracy with which the cell rate is controlled, and the timeliness of any response to violations of the contract. Let's look at some algorithms which can monitor this virtual mean.

ALGORITHMS FOR UPC

Methods to control peak cell rate, mean cell rate and different load states within several time scales have been studied extensively [11.1]. The most common algorithms involve two basic mechanisms:

- the window method, which limits the number of cells in a time window
- the leaky bucket method, which increments a counter for each cell arrival and decrements this counter periodically

The window method basically corresponds to the description given in the previous section and involves choosing a time interval and a maximum number of cells that can be admitted within that interval. We saw, with the Poisson source example, that the method suffers from either rejecting too many legitimate cells, or not rejecting enough when the contract is violated. A number of variations of the method have been studied (the jumping window, the moving window and the exponentially weighted moving average), but there is not space to deal with them here.

The leaky bucket

It is generally agreed that the leaky bucket method achieves a better performance compromise than the window method. Leaky buckets are simple to understand and to implement, and flexible in application. (Indeed, the continuous-state version of the leaky bucket algorithm is used to define the generic cell rate algorithm (GCRA), for traffic contract conformance – see [10.1, 10.2].) Figure 11.3 illustrates the principle. Note that a separate control function is required for each virtual channel or virtual path being monitored.

A counter is incremented whenever a cell arrives; this counter, which is called the 'bucket', is also decremented at a constant 'leak' rate. If the

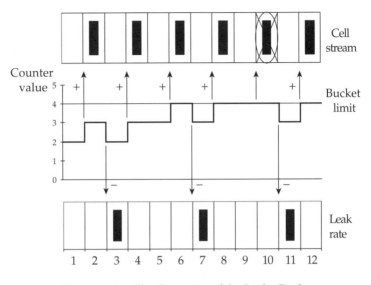

Figure 11.3. The Operation of the Leaky Bucket

traffic source generates a burst of cells at a rate higher than the leak rate, the bucket begins to fill. Provided that the burst is short, the bucket will not fill up and no action will be taken against the cell stream. If a long enough burst of cells arrives at a rate higher than the leak rate, then the bucket will eventually overflow. In this case, each cell that arrives to find the counter at its maximum value is deemed to be in violation of the traffic contract and may be discarded or 'tagged' by changing the CLP bit in the cell header from high to low priority. Another possible course of action is for the connection to be released.

In Figure 11.3, the counter has a value of 2 at the start of the sequence. The leak rate is one every four cell slots and the traffic source being monitored is in a highly active state sending cells at a rate of 50% of the cell slot rate. It is not until the tenth cell slot in the sequence that a cell arrival finds the bucket on its limit. This non-conforming cell is then subject to discard or tagging. An important point to note is that the cells do not pass through the bucket, as though queueing in a buffer. Cells do not queue in the bucket, and therefore there is no variable delay through a leaky bucket. However, the operation of the bucket can be *analysed* as though it were a buffer with cells being served at the leak rate. This then allows us to find the probability that cells will be discarded or tagged by the UPC function.

PEAK CELL RATE CONTROL USING THE LEAKY BUCKET

If life were simple, then peak cell rate control would just involve a leaky bucket with a leak rate equal to the peak rate and a bucket depth of 1. The problem is the impact of cell-delay variation (CDV), which is introduced to the cell stream by the access network. Although a source may send cells with a constant inter-arrival time at the peak rate, those cells have to go through one or more buffers in the access network before they are monitored by the UPC algorithm on entry to the public network. The effect of queueing in those buffers is to vary the amount of delay experienced by each cell. Thus the time between successive cells from the same connection may be more than or less than the declared constant inter-arrival time.

For example, suppose there are 5 CBR sources, each with a peak rate of 10% of the cell slot rate, i.e. 1 cell every 10 slots, being multiplexed through an access switch with buffer capacity of 20 cells. If all the sources are out of phase, then none of the cells suffers any queueing delay in the access switch. However, if all the sources are in phase, then the worst delay will be for the last cell in the batch, i.e. a delay of 4 cell slots (the cell which is first to arrive enters service immediately and experiences no delay). Thus the maximum variation in delay is 4 cell slots. This worst case is illustrated in Figure 11.4. At the source, the inter-arrival times between cells 1 and 2, T_{12}, and cells 2 and 3, T_{23}, are both 10 cell slots. However, cell number 2 experiences the maximum CDV of 4 cell slots, and so, on entry to the public network, the time between cells 2 and 3, T_{23}, is reduced from 10 cell slots to 6 cell slots. This corresponds to a rate increase from 10% to 16.7% of the cell slot rate, i.e. a 67% increase on the declared peak cell rate.

It is obvious that the source itself is not to blame for this apparent increase in its peak cell rate; it is just a consequence of multiplexing in the access network. However, a strict peak cell rate control, with a leak rate of 10% of the cell slot rate and a bucket limit of 1, would penalize the connection by discarding cell number 3. How is this avoided? A CDV *tolerance* is needed for the UPC function, and this is achieved by increasing the leaky bucket limit.

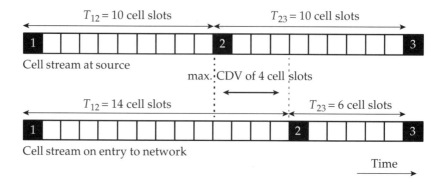

Figure 11.4.   Effect of CDV in Access Network on Inter-Arrival Times

Let's see how the leaky bucket would work in this situation. First, we must alter slightly our leaky bucket algorithm so that it can deal with any values of T (the inter-arrival time at the peak cell rate) and τ (the CDV tolerance). The leaky bucket counter works with integers, so we need to find integers k and n such that

$$\tau = k \cdot \frac{T}{n}$$

i.e. the inter-arrival time at the peak cell rate is divided into n equal parts, with n chosen so that the CDV tolerance is an integer multiple, k, of T/n. Then we operate the leaky bucket in the following way: the counter is incremented by n (the 'splash') when a cell arrives, and it is decremented at a leak rate of n/T. If the addition of a splash takes the counter above its limit of $k + n$, then the cell is in violation of the contract and is discarded or tagged. If the counter value is greater than n but less than or equal to $k + n$, then the cell is within the CDV tolerance and is allowed to enter the network.

Figure 11.5 shows how the counter value changes for the three cell arrivals of the example of Figure 11.4. In this case, $n = 10$, $k = 4$, the leak rate is equal to the cell slot rate, and the leaky bucket limit is $k + n = 14$. We assume that, when a cell arrives at the same time as the counter is decremented, the decrement takes place first, followed by the addition of the splash of n. Thus in the example shown the counter reaches, but does not exceed, its limit at the arrival of cell number 3. This is because the inter-arrival time between cells 2 and 3 has suffered the maximum CDV permitted in the traffic contract which the leaky bucket is monitoring. Figure 11.6 shows what happens for the case when cell number 2 is delayed by 5 cell slots rather than 4 cell slots. The counter exceeds its limit when cell number 3 arrives, and so that cell must be discarded because it has violated the traffic contract.

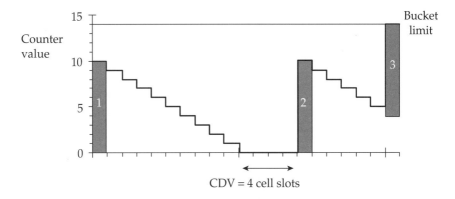

Figure 11.5.   Example of Cell Stream with CDV within the Tolerance

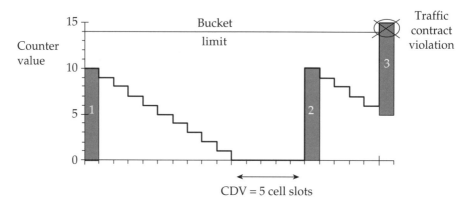

Counter
value

CDV = 5 cell slots

Figure 11.6. Example of Cell Stream with CDV Exceeding the Allowed Tolerance

The same principle applies if the tolerance, τ, exceeds the peak rate inter-arrival time, T, i.e. $k > n$. In this case it will take a number of successive cells with inter-arrival times less than T for the bucket to build up to its limit. Note that this extra parameter, the CDV tolerance, is an integral part of the traffic contract and must be specified in addition to the peak cell rate.

The problem of tolerances

When the CDV is greater than or equal to the inter-arrival time at the peak cell rate the tolerance in the UPC function presents us with a problem. It is now possible to send multiple cells at the cell slot rate. The length of this burst is limited by the size of the bucket, but if the bucket is allowed to recover, i.e. the counter returns to zero, then another burst at the cell slot rate can be sent, and so on. Thus the consequence of introducing tolerances is to allow traffic with quite different characteristics to conform to the traffic contract.

An example of this worst-case traffic is shown in Figure 11.7. The traffic contract is for a high-bandwidth (1 cell every 5 cell slots) CBR connection. With a CDV tolerance of 20 cell slots, we have $n = 1$, $k = 4$, the leak rate is the peak cell rate (20% of the cell slot rate), and the leaky bucket limit is $k + n = 5$. However, this allows a group of 6 cells to pass unhindered at the maximum cell rate of the link every 30 cell slots! So this worst-case traffic is an on/off source of the same mean cell rate but at five times the peak cell rate.

How do we calculate this maximum burst size (MBS) at the cell slot rate, and the number of empty cell slots (ECS) between such bursts? We need to analyse the operation of the leaky bucket *as though it were* a queue with cells (sometimes called 'splashes') arriving and being served. The

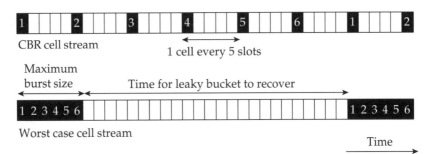

Figure 11.7. Worst-Case Traffic for Leaky Bucket with CDV Tolerance

first n units in the leaky bucket effectively act as the service space for a splash. These n units are required for the leaky bucket to operate correctly on a peak cell rate whether or not there is any CDV tolerance. Thus it is in the extra k units where a queue forms, and so the leaky bucket limit of $k + n$ is equivalent to the *system capacity*.

So, we can analyse the formation of a queue by considering the time taken, t_{MBS}, for an excess rate to fill up the leaky bucket's *queueing space*, k:

$$t_{MBS} = \frac{\text{queueing space}}{\text{excess rate}} = \frac{k}{n \cdot \text{CSR} - n \cdot \text{PCR}}$$

where CSR is the cell slot rate, and PCR is the peak cell rate. We also know that

$$k = \tau \cdot \frac{n}{T} = \tau \cdot n \cdot \text{PCR}$$

so, substituting for k gives

$$t_{MBS} = \frac{\tau \cdot n \cdot \text{PCR}}{n \cdot (\text{CSR} - \text{PCR})} = \frac{\tau \cdot \text{PCR}}{\text{CSR} - \text{PCR}}$$

The maximum burst size is found by multiplying t_{MBS} by the cell slot rate and adding one for the first cell in the burst which fills the server space, n:

$$\text{MBS} = 1 + \lfloor \text{CSR} \cdot t_{MBS} \rfloor = 1 + \left\lfloor \frac{\tau \cdot \text{CSR} \cdot \text{PCR}}{\text{CSR} - \text{PCR}} \right\rfloor$$

which, in terms of the inter-arrival time at the peak rate, T, and the cell slot time, Δ, is

$$\text{MBS} = 1 + \left\lfloor \frac{\tau}{T - \Delta} \right\rfloor$$

We take the integer part of this calculation because there cannot be fractions of cells in a burst. For the example given in Figure 11.7, we have

$$MBS = 1 + \left\lfloor \frac{20}{5-1} \right\rfloor = 6 \text{ cells}$$

The time taken, t_{cycle}, for the leaky bucket to go through one complete cycle of filling, during the maximum burst, and then emptying during the silence period, is given by

$$n \cdot MBS = n \cdot PCR \cdot t_{cycle}$$

where the left-hand side gives the total number of units by which the leaky bucket is incremented, and the right-hand side gives the total number by which it is decremented. The total number of cell slots in a complete cycle is $CSR \cdot t_{cycle}$. It is necessary to round this up to the nearest integer number of cell slots to ensure that the leaky bucket empties completely, so the number of empty cell slots is

$$ECS = \left\lceil CSR \cdot \frac{MBS}{PCR} \right\rceil - MBS$$

For the example given in Figure 11.7, we have

$$ECS = \left\lceil 1 \cdot \frac{6}{0.2} \right\rceil - 6 = 24 \text{ cells}$$

Resources required for a worst-case ON/OFF cell stream from peak cell rate UPC

Continuing with this example, suppose that there are five of these CBR sources each being controlled by its own leaky bucket with the parameter values calculated. After the UPC function, the cell streams are multiplexed through an ATM buffer of capacity 20 cells. If the sources do in fact behave according to their declared contract, i.e. a peak cell rate of 20% of the cell slot rate, then there is no cell loss. In any five cell slots, we know that there will be exactly five cells arriving; this can be accommodated in the ATM buffer without loss.

But if all five sources behave as the worst-case ON/OFF cell stream, then the situation is different. We know that in any 30 cell slots there will be exactly 30 cells arriving. Whether the buffer capacity of 20 cells is sufficient depends on the relative phasing of the ON/OFF cell streams. If all five sources send a maximum burst at the same time, then 30 cells will arrive during six cell slots. This is an excess of 24 cells, 20 of which can be buffered, and 4 of which must be lost.

If we reduce the number of sources to four, their worst-case behaviour will produce an aggregate burst of 24 cells arriving during six cell slots. This is an excess of 18 cells, all of which can be buffered. Thus the performance can be maintained only if the number of sources, and hence the admissible load, is reduced.

The example we have used is a very simple one that demonstrates the issue. A reduction from five sources admitted to four may not seem to be a severe consequence of CDV tolerances. In general, each CBR source of peak cell rate h cell/s is, in the worst case, being considered as an ON/OFF source with a *mean* cell rate of h cell/s and a peak cell rate equal to the cell slot rate of the link. This can reduce the admissible load by a significant amount.

We can estimate this load reduction by applying the N·D/D/1 analysis to the worst-case traffic streams. The application of this analysis rests on the observation that the worst-case ON/OFF source is in fact periodic, with period MBS · D. Each arrival is a burst of fixed size, MBS, which takes MBS cell slots to be served, so the period can be described as one burst in every D burst slots. A buffer of size X cells can hold X/MBS bursts, so we can adapt the analysis to deal with bursts rather than cells just by scaling the buffer capacity. The analysis does not account for the partial overlapping of bursts, but since we are only after an estimate we will neglect this detail.

The approximate analysis for the N·D/D/1 queue tends to underestimate the loss particularly when the load is not heavy. The effect of this is to overestimate the admissible load for a fixed CLP requirement. But the great advantage is that we can manipulate the formula to give the admissible load, ρ, as a function of the other parameters, X and D:

$$\rho = \frac{2 \cdot (X + D)}{D \cdot \left(2 - \dfrac{\ln(\text{CLP})}{X}\right)}$$

with the proviso that the load can never exceed a value of 1. This formula applies to the CBR cell streams. For the worst-case streams, we just replace X by X/MBS to give:

$$\rho = \frac{2 \cdot \left(\dfrac{X}{\text{MBS}} + D\right)}{D \cdot \left(2 - \dfrac{\text{MBS} \cdot \ln(\text{CLP})}{X}\right)}$$

where

$$\text{MBS} = 1 + \left\lfloor \frac{\tau}{T - \Delta} \right\rfloor = 1 + \left\lfloor \frac{\tau/\Delta}{D - 1} \right\rfloor$$

Note that D is just the inter-arrival time, T, in units of the cell slot time, Δ.

Table 11.2 shows the number of sources ($N = \rho \cdot D$) that can be admitted for different CDV values and with a CLP requirement of 10^{-10}. It is assumed that the output cell streams from N UPC functions are multiplexed together over a 155.52 Mbit/s link (i.e. a cell slot rate of 353 208 cell/s). The link buffer has a capacity of 50 cells. The CDV tolerance allowed by the leaky buckets takes values of 20, 40, 60, 80 and 100 cell

Table 11.2. Number of CBR Sources that can Be Admitted over a 155.52 Mbit/s Link with Buffer Capacity of 50 Cells, for Different CDV Tolerances and a CLP of 10^{-10}

period, D (slots)	cell rate (cell/s)	Cell delay variation tolerance					
		$\tau/\Delta = 0$ slots $\tau = 0$ ms	20 slots 0.057 ms	40 slots 0.113 ms	60 slots 0.170 ms	80 slots 0.226 ms	100 slots 0.283 ms
10	35 321	10	10	9	6	5	3
11	32 110	11	11	9	6	5	4
12	29 434	12	12	12	8	6	5
13	27 170	13	13	13	8	7	6
14	25 229	14	14	13	11	8	7
15	23 547	15	15	15	11	9	7
16	22 075	16	16	16	12	10	8
17	20 777	17	17	17	15	10	9
18	19 623	18	18	18	15	13	11
19	18 590	19	19	19	16	13	11
20	17 660	20	20	20	16	13	11
21	16 819	21	21	21	17	14	12
22	16 055	22	22	22	22	17	14
23	15 357	23	23	23	23	18	15
24	14 717	24	24	24	24	19	15
25	14 128	25	25	25	24	19	16
26	13 585	26	26	26	25	20	16
27	13 082	27	27	27	25	20	20
28	12 615	28	28	28	26	26	21
29	12 180	29	29	29	27	27	21
30	11 774	30	30	30	27	27	22
35	10 092	35	35	35	35	30	30
40	8 830	40	40	40	40	33	33
45	7 849	45	45	45	45	45	36
50	7 064	50	50	50	50	50	39
55	6 422	55	55	55	54	54	54
60	5 887	60	60	60	58	58	58
65	5 434	65	65	65	65	61	61
70	5 046	70	70	70	70	65	65
75	4 709	75	75	75	75	68	68
80	4 415	80	80	80	80	71	71
85	4 155	85	85	85	85	85	75
90	3 925	90	90	90	90	90	78
95	3 718	95	95	95	95	95	82
100	3 532	100	100	100	100	100	85

slots (corresponding to time values of 0.057, 0.113, 0.17, 0.226 and 0.283 ms respectively). The peak cell rates being monitored vary from 1% up to 10% of the cell slot rate. If the CDV tolerance is zero, then in this case the link can be loaded to 100% of capacity for each of the peak cell rates shown.

Figure 11.8 plots the data of Table 11.2 as the admissible load against the monitored peak cell rate. Note that when the CDV is relatively small (e.g. 40 cell slots or less), then there is little or no reduction in the admissible

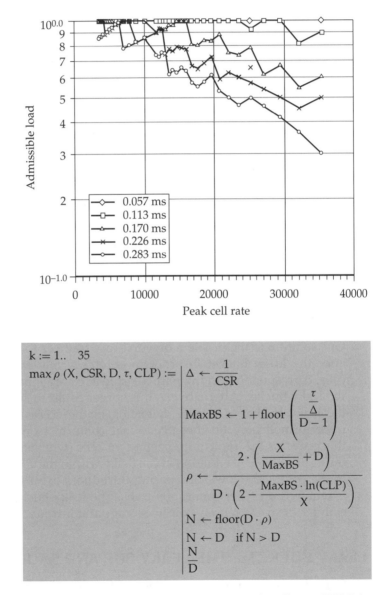

Figure 11.8. Admissible Load for CBR Sources with Different CDV Tolerances

$$D_k := \begin{vmatrix} (k+9) & \text{if } k < 22 \\ (k-21) \cdot 5 + 30 & \text{otherwise} \end{vmatrix}$$

$$CSR := 353207.5$$

$$PCR_k := \frac{CSR}{D_k}$$

$$CLP := 10^{-10}$$

$$y1_k := \max \rho \left(50, CSR, D_k, \frac{20}{CSR}, CLP \right)$$

$$y2_k := \max \rho \left(50, CSR, D_k, \frac{40}{CSR}, CLP \right)$$

$$y3_k := \max \rho \left(50, CSR, D_k, \frac{60}{CSR}, CLP \right)$$

$$y4_k := \max \rho \left(50, CSR, D_k, \frac{80}{CSR}, CLP \right)$$

$$y5_k := \max \rho \left(50, CSR, D_k, \frac{100}{CSR}, CLP \right)$$

Figure 11.8.   (*continued*)

load in this example. The CDV in the access network may well be of this order, particularly if the access network utilization is low and buffers are dimensioned to cope with only cell-scale and not burst-scale queueing.

Traffic shaping

One solution to the problem of worst-case traffic is to introduce a spacer after the leaky bucket in order to enforce a minimum time between cells, corresponding to the particular peak cell-rate being monitored by the leaky bucket. Alternatively, this spacer could be implemented *before* the leaky bucket as per-VC queueing in the access network. Spacing is performed only on those cells that conform to the traffic contract; this prevents the 'bunching together' of cells (whether of the worst-case traffic or caused by variation in cell delay within the CDV tolerance of the traffic contract). However, spacing introduces extra complexity, which is required on a per-connection basis. The leaky bucket is just a simple counter—a spacer requires buffer storage and introduces delay.

DUAL LEAKY BUCKETS: THE LEAKY CUP AND SAUCER

Consider the situation for a variable-rate source described by a peak cell rate and a mean cell rate. This can be monitored by two leaky buckets:

one to control the peak cell rate, the other to control a 'virtual mean' cell rate. In ITU Recommendation I.371 the term used for this 'virtual mean' is 'sustainable cell rate' (SCR). With two leaky buckets, the effect of the CDV tolerance on the peak-cell-rate leaky bucket is not so severe. The reason is that the leaky bucket for the sustainable cell rate limits the number of worst-case bursts that can pass through the peak-cell-rate leaky bucket. For each ON/OFF cycle at the cell slot rate the SCR leaky-bucket level increases by a certain amount. When the SCR leaky bucket reaches its limit, the ON/OFF cycles must stop until the SCR counter has returned to zero. So the maximum burst size is still determined by the PCR leaky-bucket parameter values, but the overall mean cell rate allowed onto the network is limited by the sustainable cell rate rather than the peak cell rate.

This dual leaky bucket arrangement is called the 'leaky cup and saucer'. The cup is the leaky bucket for the sustainable cell rate: it is a deep container with a base of relatively small diameter. The saucer is the leaky bucket for the peak cell rate: it is a shallow container with a large-diameter base. The depth corresponds to the bucket limit and the diameter of the base to the cell rate being controlled.

The worst-case traffic is shown in Figure 11.9(a). The effect of the leaky buckets is to limit the number of cells over different time periods. For the example in the figure, the saucer limit is 2 cells in 4 cell slots and the cup limit is 6 cells in 24 cell slots. An alternative 'worst-case' traffic which is adopted in ITU Recommendation E.736 [10.3] is an ON/OFF source with maximum-length bursts at the peak cell rate rather than at the cell slot rate. An example of this type of worst-case traffic is shown in Figure 11.9(b). Note that the time axis is in cell slots, so the area under the curve is equal to the number of cells sent.

The maximum burst size at the peak cell rate is obtained in a similar way to that at the cell slot rate, i.e.

$$\text{MBS} = 1 + \left\lfloor \frac{\tau_{\text{IBT}}}{T_{\text{SCR}} - T_{\text{PCR}}} \right\rfloor$$

where τ_{IBT} is called the 'intrinsic burst tolerance'. This is another important parameter in the traffic contract (in addition to the inter-arrival times T_{SCR} and T_{PCR} for the sustainable and peak cell rates respectively). The purpose of the intrinsic burst tolerance is in fact to specify the burst length limit in the traffic contract.

Two CDV tolerances are specified in the traffic contract. We are already familiar with the CDV tolerance, τ, for the peak cell rate. From now on we call this τ_{PCR} to distinguish it from the CDV tolerance for the sustainable cell rate, τ'_{SCR}. This latter has to be added to the intrinsic burst tolerance in order to determine the counter limit for the cup. As before, we need to

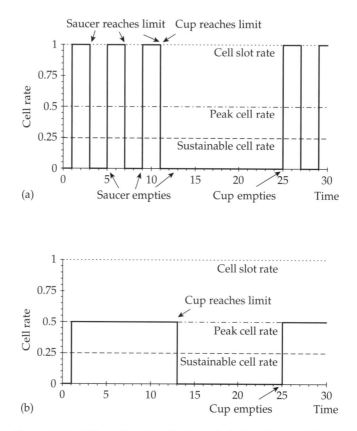

Figure 11.9. Worst-Case Traffic through Leaky Cup and Saucer

find integers, k and n, such that

$$\tau_{IBT} + \tau'_{SCR} = k \cdot \frac{T_{SCR}}{n}$$

In most cases, n can be set to 1 because the intrinsic burst tolerance will be many times larger than T_{SCR}.

Resources required for a worst-case ON/OFF cell stream from sustainable-cell-rate UPC

Neither type of 'worst-case' traffic shown in Figure 11.9 easy to analyse. In the following analysis we use the maximum burst size for the sustainable cell rate, and assume that that burst actually arrives at the cell slot rate. Whether or not this is possible depends on the size of the saucer, and hence on τ_{PCR}. It is likely to be the worst of all possible traffic streams because it generates the largest burst size.

The same approximate analytical approach is taken as before. In this case D is the inter-arrival time, T_{SCR}, in units of the cell slot time, Δ.

$$\rho = \frac{2 \cdot \left(\dfrac{X}{MBS} + D\right)}{D \cdot \left(2 - \dfrac{MBS \cdot \ln(CLP)}{X}\right)}$$

A graph of utilization against the maximum burst size is shown in Figure 11.10. The CLP requirement varies from 10^{-4} down to 10^{-10}. The

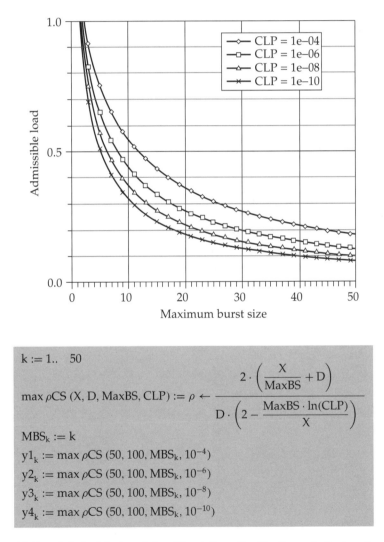

Figure 11.10.   Admissible Load for Worst-Case Traffic through Leaky Cup and Saucer

link buffer has a capacity of 50 cells, the cell slot rate is 353 208 cell/s, and the sustainable cell rate is chosen to be 3532 cell/s, i.e. $D = 100$. The maximum burst size allowed by the leaky cup and saucer is varied from 1 up to 50 cells. The peak cell rate and intrinsic burst tolerance are not specified explicitly; different combinations can be calculated from the maximum burst size and the sustainable cell rate.

It is important to use the correct value of MBS because this obviously can have a significant effect on the admissible load. Suppose that the peak cell rate is twice the sustainable cell rate, i.e. $T_{PCR} = T_{SCR}/2$. The maximum burst size at the peak cell rate is

$$
\mathrm{MBS}_{PCR} = 1 + \left\lfloor \frac{\tau_{IBT}}{T_{SCR} - \dfrac{T_{SCR}}{2}} \right\rfloor = 1 + \left\lfloor \frac{2 \cdot \tau_{IBT}}{T_{SCR}} \right\rfloor
$$

and the maximum burst size at the cell slot rate is

$$
\mathrm{MBS}_{CSR} = 1 + \left\lfloor \frac{\tau_{IBT}}{T_{SCR} - \Delta} \right\rfloor \approx 1 + \left\lfloor \frac{\tau_{IBT}}{T_{SCR}} \right\rfloor
$$

The difference between these two maximum size bursts is almost a factor of two (for reasonable values of the intrinsic burst tolerance), and this corresponds to a difference in the admissible load of a factor of roughly 0.6 across the range of burst sizes in the graph. So the assumption that the worst-case traffic is based on the maximum burst size at the peak cell rate carries with it a 40% penalty on the admissible load.

12 **Dimensioning**

real networks don't lose cells?

COMBINING THE BURST AND CELL SCALES

The finite-capacity buffer is a fundamental element of ATM where cells multiplexed from a number of different input streams are temporarily stored awaiting onward transmission. The flow of cells from the different inputs, the number of inputs, and the rate at which cells are served determine the occupancy of the buffer and hence the cell delay and cell loss experienced. So, how large should this finite buffer be?

In Chapters 8 and 9 we have seen that there are two elements of queueing behaviour: the cell-scale and burst-scale components. We evaluated the loss from a finite buffer for constant bit-rate, variable bit-rate and random traffic sources. For random traffic, or for a mix of CBR traffic, only the cell-scale component is present. But when the traffic mix includes bursty sources, such that combinations of the active states can exceed the cell slot rate, then both components of queueing are present.

Let's look at each type of traffic and see how the loss varies with the buffer size for different offered loads. We can then develop strategies for buffer dimensioning based on an understanding of this behaviour. First, we consider VBR traffic; this combines the cell-scale component of queueing with both the loss and delay factors of the burst-scale component of queueing.

Figure 9.14 shows how the burst-scale loss factor varies with the number of sources, N, where each source has a peak cell rate of 24 000 cell/s and a mean cell rate of 2000 cell/s. From Table 9.2 we find that the minimum number of these sources required for burst-scale queueing is $N_0 = 14.72$. Table 12.1 gives the burst-scale loss factor, CLP_{bsl}, at three different values of N (30, 60 and 90 sources) as well as the offered load as a fraction of the cell slot rate (calculated using the bufferless analysis in Chapter 9). These values of load are used to calculate both the

Table 12.1. Burst-Scale Loss Factor for N VBR Sources

N	CLP_{bsl}	load
30	4.46E-10	0.17
60	1.11E-05	0.34
90	9.10E-04	0.51

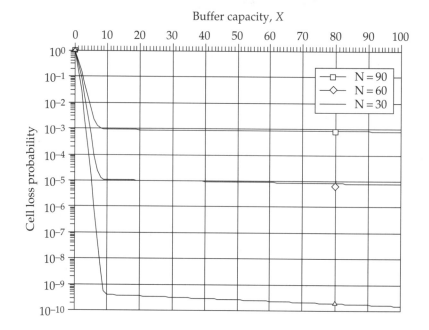

```
k := 2.. 100
OverallCLP (X, N, m, h, C, b) := | ρ ← N·m/C
                                  |
                                  | N0 ← C/h
                                  |
                                  | α ← m/h
                                  | for i ∈ 0.. X
                                  | aᵢ ← Poisson (i, ρ)
                                  | csloss ← finiteQloss (X, a, ρ)
                                  | bsloss ← BSLexact (α, N, N0) · BSDapprox (N0, X, b, ρ)
                                  | csloss + bsloss

xₖ := k
y1ₖ := OverallCLP (k , 90 , 2000 , 24000 , 353207.5 , 480 )
y2ₖ := OverallCLP (k , 60 , 2000 , 24000 , 353207.5 , 480 )
y3ₖ := OverallCLP (k , 30 , 2000 , 24000 , 353207.5 , 480 )
```

Figure 12.1. Overall Cell Loss Probability against Buffer Capacity for N VBR Sources

cell-scale queueing component, CLP_{cs}, and the burst-scale delay factor, CLP_{bsd}, varying with buffer capacity.

The combined results are plotted in Figure 12.1. The cell-scale component is obtained using the exact analysis of the finite M/D/1 described in Chapter 7. The burst-scale delay factor uses the same approach as that for calculating the values in Figure 9.16. For Figure 12.1, an average burst length, b, of 480 cells is used. The overall cell loss shown in Figure 12.1 is calculated by summing the burst- and cell-scale components of cell loss, where the burst-scale component is the product of the loss and delay factors, i.e.

$$CLP = CLP_{cs} + CLP_{bsl} \cdot CLP_{bsd}$$

Now, consider N CBR sources where each source has a constant cell rate of 2000 cell/s. Figure 12.2 shows how the cell loss varies with the buffer

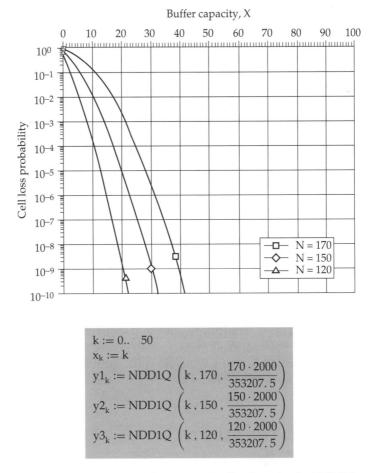

Figure 12.2. Cell Loss Probability against Buffer Capacity for N CBR Sources

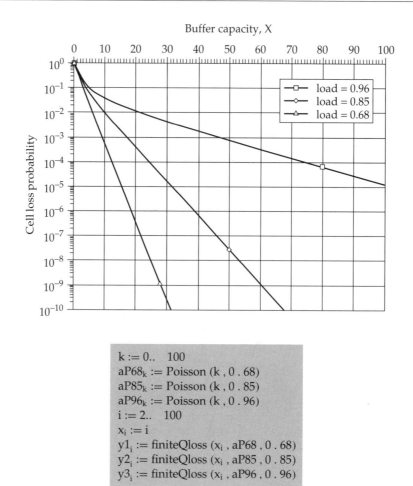

Buffer capacity, X

Figure 12.3. Cell Loss Probability against Buffer Capacity for Random Traffic

capacity for 120, 150 and 170 sources. The corresponding values for the offered load are 0.68, 0.85, and 0.96 respectively. Figure 12.3 takes the load values used for the CBR traffic and assumes that the traffic is random. The cell loss results are found using the exact analysis for the finite M/D/1 system. A summary of the three different situations is depicted in Figure 12.4, comparing 30 VBR sources, 150 CBR sources, and an offered load of 0.85 of random traffic (the same load as 150 CBR sources).

DIMENSIONING THE BUFFER

Figure 12.4 shows three very different curves, depending on the characteristics of each different type of source. There is no question that the

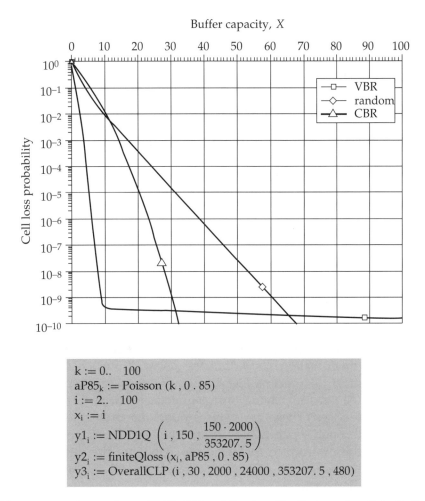

Figure 12.4. Comparison of VBR, CBR and Random Traffic through a Finite Buffer

buffer must be able to cope with the cell-scale component of queueing since this is always present when a number of traffic streams are merged. But we have two options when it comes to the burst-scale component, as analysed in Chapter 9:

1. Restrict the number of bursty sources so that the total input rate only rarely exceeds the cell slot rate, and assume that all excess-rate cells are lost. This is the bufferless or burst-scale loss option (also known as 'rate envelope multiplexing').

2. Assume that we have a big enough buffer to cope with excess-rate cells, so only a proportion are lost; the other excess-rate cells are delayed in the buffer. This is the burst-scale delay option (rate-sharing statistical multiplexing).

It is important to notice that how big we make the buffer depends on how we intend to accept traffic onto the network (or vice versa). Also a dimensioning choice has an impact on a control mechanism (connection admission control).

For the first option, the buffer is dimensioned according to cell-scale constraints. The amount of bursty traffic is not the limiting factor in choosing the buffer capacity because the CAC restrictions on accepting bursty traffic automatically limit the burst-scale component to a value below the CLP requirement, and the CAC algorithm assumes that the buffer size makes no difference. Thus for bursty traffic the mean utilization is low and the gradient of its cell-scale component is steep (see Figure 12.1). However, for either constant-bit-rate or random traffic the cell-scale component is much more significant (there is no burst-scale component), and it is a realistic maximum load of this traffic that determines the buffer capacity. The limiting factor here is the delay through the buffer, particularly for interactive services.

If we choose the second option, the amount of bursty traffic can be increased to the same levels of utilization as for either constant-bit-rate or random traffic – the price to pay is in the size of the buffer which must be significantly larger. The disadvantage with buffering the excess (burst-scale) cells is that the delay through a *large* buffer can be too great for services like telephony and interactive video, which negates the aims of having an integrated approach to all telecommunications services. There are ways around the problem – segregation of traffic through separate buffers and the use of time priority servers – but this does introduce further complexity into the network, see Figure 12.5. We will look at traffic segregation and priorities in more detail in Chapter 13.

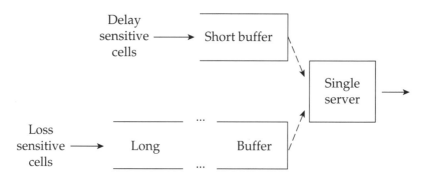

A time priority scheme would involve serving cells
in the short buffer before cells in the long buffer

Figure 12.5.   Time Priorities and Segregation of Traffic

Small buffers for cell-scale queueing

A comparison of random traffic and CBR traffic (see Figure 12.4) shows that the 'cell-scale component' of the random traffic gives a worse CLP for the same load. Even with 1000 CBR sources, each of 300 cell/s (to keep the load constant at 0.85), Table 10.3(b) shows that the cell loss is about 10^{-9} for a buffer capacity of 50 cells. This is a factor of 10 lower than for random traffic through the same size buffer.

So, to dimension buffers for cell-scale queueing we use a realistic maximum load of *random* traffic. Table 12.2 uses the exact analysis for

Table 12.2

load	buffer capacity (cells)	155.52 Mbit/s link		622.08 Mbit/s link	
		mean delay (µs)	maximum delay (µs)	mean delay (µs)	maximum delay (µs)

(a) Buffer Dimensioning for Cell-Scale Queueing: Buffer Capacity, Mean and Maximum Delay, Given the Offered Load and a Cell Loss Probability of 10^{-8}

load	buffer capacity (cells)	mean delay (µs)	maximum delay (µs)	mean delay (µs)	maximum delay (µs)
0.50	16	4.2	45.3	1.1	11.3
0.51	16	4.3	45.3	1.1	11.3
0.52	17	4.4	48.1	1.1	12.0
0.53	17	4.4	48.1	1.1	12.0
0.54	17	4.5	48.1	1.1	12.0
0.55	18	4.6	51.0	1.1	12.7
0.56	18	4.6	51.0	1.2	12.7
0.57	19	4.7	53.8	1.2	13.4
0.58	19	4.8	53.8	1.2	13.4
0.59	20	4.9	56.6	1.2	14.2
0.60	20	5.0	56.6	1.2	14.2
0.61	21	5.0	59.5	1.3	14.9
0.62	21	5.1	59.5	1.3	14.9
0.63	22	5.2	62.3	1.3	15.6
0.64	23	5.3	65.1	1.3	16.3
0.65	23	5.5	65.1	1.4	16.3
0.66	24	5.6	67.9	1.4	17.0
0.67	25	5.7	70.8	1.4	17.7
0.68	25	5.8	70.8	1.5	17.7
0.69	26	6.0	73.6	1.5	18.4
0.70	27	6.1	76.4	1.5	19.1
0.71	28	6.3	79.3	1.6	19.8
0.72	29	6.5	82.1	1.6	20.5
0.73	30	6.7	84.9	1.7	21.2
0.74	31	6.9	87.8	1.7	21.9
0.75	33	7.1	93.4	1.8	23.4
0.76	34	7.3	96.3	1.8	24.1
0.77	35	7.6	99.1	1.9	24.8

(continued overleaf)

Table 12.2.   (*continued*)

load	buffer capacity (cells)	155.52 Mbit/s link		622.08 Mbit/s link	
		mean delay (µs)	maximum delay (µs)	mean delay (µs)	maximum delay (µs)
0.78	37	7.9	104.8	2.0	26.2
0.79	39	8.2	110.4	2.0	27.6
0.80	41	8.5	116.1	2.1	29.0
0.81	43	8.9	121.7	2.2	30.4
0.82	45	9.3	127.4	2.3	31.9
0.83	48	9.7	135.9	2.4	34.0
0.84	51	10.3	144.4	2.6	36.1
0.85	54	10.9	152.9	2.7	38.2
0.86	58	11.5	164.2	2.9	41.1
0.87	62	12.3	175.5	3.1	43.9
0.88	67	13.2	189.7	3.3	47.4
0.89	73	14.3	206.7	3.6	51.7
0.90	79	15.6	223.7	3.9	55.9
0.91	88	17.1	249.1	4.3	62.3
0.92	98	19.1	277.5	4.8	69.4
0.93	112	21.6	317.1	5.4	79.3
0.94	129	25.0	365.2	6.3	91.3
0.95	153	29.7	433.2	7.4	108.3
0.96	189	36.8	535.1	9.2	133.8
0.97	248	48.6	702.1	12.2	175.5
0.98	362	72.2	1024.9	18.0	256.2

(b) Buffer Dimensioning for Cell-Scale Queueing: Buffer Capacity, Mean and Maximum Delay, Given the Offered Load and a Cell Loss Probability of 10^{-10}

load	buffer capacity (cells)	155.52 Mbit/s link		622.08 Mbit/s link	
0.50	19	4.2	53.8	1.1	13.4
0.51	20	4.3	56.6	1.1	14.2
0.52	20	4.4	56.6	1.1	14.2
0.53	21	4.4	59.5	1.1	14.9
0.54	21	4.5	59.5	1.1	14.9
0.55	22	4.6	62.3	1.1	15.6
0.56	23	4.6	65.1	1.2	16.3
0.57	23	4.7	65.1	1.2	16.3
0.58	24	4.8	67.9	1.2	17.0
0.59	24	4.9	67.9	1.2	17.0
0.60	25	5.0	70.8	1.2	17.7
0.61	26	5.0	73.6	1.3	18.4
0.62	26	5.1	73.6	1.3	18.4
0.63	27	5.2	76.4	1.3	19.1
0.64	28	5.3	79.3	1.3	19.8
0.65	29	5.5	82.1	1.4	20.5
0.66	30	5.6	84.9	1.4	21.2
0.67	31	5.7	87.8	1.4	21.9
0.68	32	5.8	90.6	1.5	22.6
0.69	33	6.0	93.4	1.5	23.4

Table 12.2. *(continued)*

load	buffer capacity (cells)	155.52 Mbit/s link		622.08 Mbit/s link	
		mean delay (µs)	maximum delay (µs)	mean delay (µs)	maximum delay (µs)
0.70	34	6.1	96.3	1.5	24.1
0.71	35	6.3	99.1	1.6	24.8
0.72	37	6.5	104.8	1.6	26.2
0.73	38	6.7	107.6	1.7	26.9
0.74	39	6.9	110.4	1.7	27.6
0.75	41	7.1	116.1	1.8	29.0
0.76	43	7.3	121.7	1.8	30.4
0.77	45	7.6	127.4	1.9	31.9
0.78	47	7.9	133.1	2.0	33.3
0.79	49	8.2	138.7	2.0	34.7
0.80	51	8.5	144.4	2.1	36.1
0.81	54	8.9	152.9	2.2	38.2
0.82	57	9.3	161.4	2.3	40.3
0.83	60	9.7	169.9	2.4	42.5
0.84	64	10.3	181.2	2.6	45.3
0.85	68	10.9	192.5	2.7	48.1
0.86	73	11.5	206.7	2.9	51.7
0.87	79	12.3	223.7	3.1	55.9
0.88	85	13.2	240.7	3.3	60.2
0.89	93	14.3	263.3	3.6	65.8
0.90	102	15.6	288.8	3.9	72.2
0.91	113	17.1	319.9	4.3	80.0
0.92	126	19.1	356.7	4.8	89.2
0.93	144	21.6	407.7	5.4	101.9
0.94	167	25.0	472.8	6.3	118.2
0.95	199	29.7	563.4	7.4	140.9
0.96	246	36.8	696.5	9.2	174.1
0.97	324	48.6	917.3	12.2	229.3
0.98	476	72.2	1347.6	18.0	336.9

(c) Buffer Dimensioning for Cell-Scale Queueing: Buffer Capacity, Mean and Maximum Delay, Given the Offered Load and a Cell Loss Probability of 10^{-12}

load	buffer capacity (cells)	mean delay (µs)	maximum delay (µs)	mean delay (µs)	maximum delay (µs)
0.50	23	4.2	65.1	1.1	16.3
0.51	24	4.3	67.9	1.1	17.0
0.52	24	4.4	67.9	1.1	17.0
0.53	25	4.4	70.8	1.1	17.7
0.54	26	4.5	73.6	1.1	18.4
0.55	26	4.6	73.6	1.1	18.4
0.56	27	4.6	76.4	1.2	19.1
0.57	28	4.7	79.3	1.2	19.8
0.58	28	4.8	79.3	1.2	19.8
0.59	29	4.9	82.1	1.2	20.5
0.60	30	5.0	84.9	1.2	21.2
0.61	31	5.0	87.8	1.3	21.9

(continued overleaf)

Table 12.2. (*continued*)

load	buffer capacity (cells)	155.52 Mbit/s link		622.08 Mbit/s link	
		mean delay (μs)	maximum delay (μs)	mean delay (μs)	maximum delay (μs)
0.62	32	5.1	90.6	1.3	22.6
0.63	33	5.2	93.4	1.3	23.4
0.64	34	5.3	96.3	1.3	24.1
0.65	35	5.5	99.1	1.4	24.8
0.66	36	5.6	101.9	1.4	25.5
0.67	37	5.7	104.8	1.4	26.2
0.68	38	5.8	107.6	1.5	26.9
0.69	39	6.0	110.4	1.5	27.6
0.70	41	6.1	116.1	1.5	29.0
0.71	42	6.3	118.9	1.6	29.7
0.72	44	6.5	124.6	1.6	31.1
0.73	46	6.7	130.2	1.7	32.6
0.74	47	6.9	133.1	1.7	33.3
0.75	49	7.1	138.7	1.8	34.7
0.76	52	7.3	147.2	1.8	36.8
0.77	54	7.6	152.9	1.9	38.2
0.78	56	7.9	158.5	2.0	39.6
0.79	59	8.2	167.0	2.0	41.8
0.80	62	8.5	175.5	2.1	43.9
0.81	65	8.9	184.0	2.2	46.0
0.82	69	9.3	195.4	2.3	48.8
0.83	73	9.7	206.7	2.4	51.7
0.84	78	10.3	220.8	2.6	55.2
0.85	83	10.9	235.0	2.7	58.7
0.86	89	11.5	252.0	2.9	63.0
0.87	96	12.3	271.8	3.1	67.9
0.88	104	13.2	294.4	3.3	73.6
0.89	113	14.3	319.9	3.6	80.0
0.90	124	15.6	351.1	3.9	87.8
0.91	138	17.1	390.7	4.3	97.7
0.92	154	19.1	436.0	4.8	109.0
0.93	176	21.6	498.3	5.4	124.6
0.94	204	25.0	577.6	6.3	144.4
0.95	244	29.7	690.8	7.4	172.7
0.96	303	36.8	857.9	9.2	214.5
0.97	400	48.6	1132.5	12.2	283.1
0.98	592	72.2	1676.1	18.0	419.0

the finite M/D/1 queue to show the buffer capacity for a given load and cell loss probability. The first column is the load, varying from 50% up to 98%, and the second column gives the buffer size for a particular cell loss probability requirement (Table 12.2 part (a) is for a CLP of 10^{-8}, part (b) is for 10^{-10}, and part (c) is for 10^{-12}). Then there are extra columns which give the mean delay and maximum delay through the buffer for link rates of 155.52 Mbit/s and 622.08 Mbit/s. The maximum delay is just the buffer capacity multiplied by the time per cell slot, s, at the appropriate link rate. The mean delay depends on the load, ρ, and is calculated using the formula for an infinite M/D/1 system:

$$t_q = s + \frac{\rho \cdot s}{2 \cdot (1 - \rho)}$$

This is very close to the mean delay through a finite M/D/1 because the loss is extremely low (mean delays only differ noticeably when the loss from the finite system is high).

Figure 12.6 presents the mean and maximum delay values from Table 12.2 in the form of a graph and clearly shows how the delays increase substantially above a load of about 80%. This graph can be used to select a maximum load according to the cell loss and delay constraints, and the buffer's link rate; the required buffer size can then be read from the appropriate part of Table 12.2.

So, to summarize, we dimension short buffers to cope with cell-scale queueing behaviour using Table 12.2 and Figure 12.6. This approach is applicable to networks which offer the deterministic bit-rate transfer capability and the statistical bit-rate transfer capability based on rate envelope multiplexing. For SBR based on rate sharing, buffer dimensioning requires a different approach, based on the burst-scale queueing behaviour.

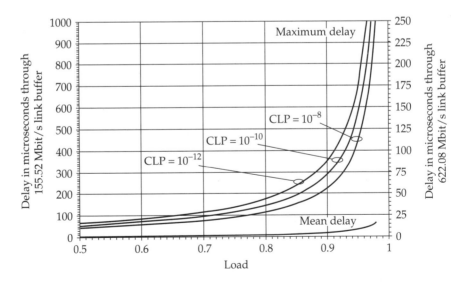

Figure 12.6. Mean and Maximum Delays for a Buffer with Link Rate of either 155.52 Mbit/s or 622.08 Mbit/s for Cell Loss Probabilities of 10^{-8}, 10^{-10} and 10^{-12}

Large buffers for burst-scale queueing

A buffer-dimensioning method for large buffers and burst-scale queueing is rather more complicated than for short buffers and cell-scale queueing because the traffic characterization has more parameters. For the cell-scale queueing case, random traffic is a very good upper bound and it has just the one parameter: arrival rate. In the burst-scale queueing case, we must assume a traffic mix of many on/off sources, each source having the same traffic characteristics (peak cell rate, mean cell rate, and the mean burst length in the active state). For the burst-scale analytical approach we described in Chapter 9, the key parameters are the minimum number of peak rates required for burst-scale queueing, N_0, the ratio of buffer capacity to mean burst length, X/b, the mean load, ρ, and the cell loss probability.

We have seen in Chapter 10 that the overall cell loss target can be obtained by trial and error with tables; combining the burst-scale loss and burst-scale delay factors from Table 10.4 and Table 10.5 respectively. Here, we present buffer-dimensioning data in two alternative graphical forms: the variation of X/b with load for a fixed value of N_0 and different overall CLP values (Figure 12.7); and the variation of X/b with load for a fixed overall CLP value and different values of N_0 (Figure 12.8).

To produce a graph like that of Figure 12.7, we take one row from Table 10.4, for a particular value of N_0. This gives the cell loss contribution, CLP_{bsl}, from the burst-scale loss factor varying with offered load

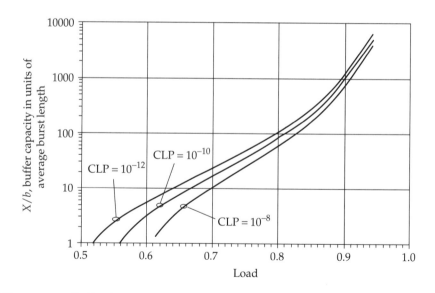

Figure 12.7. Buffer Capacity in Units of Mean Burst Length, Given Load and Cell Loss Probability, for Traffic with a Peak Cell Rate of 1/100 of the Cell Slot Rate (i.e. $N_0 = 100$)

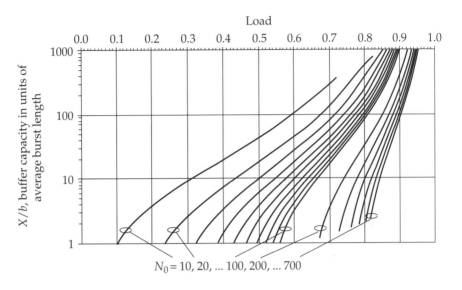

Figure 12.8. Buffer Capacity in Units of Mean Burst Length, Given Load and the Ratio of Cell Slot Rate to Peak Cell Rate, for a Cell Loss Probability of 10^{-10}

(Table 12.3). We then need to find the value of X/b which gives a cell loss contribution, CLP_{bsd}, from the burst-scale delay factor, to meet the overall CLP target. This is found by rearranging the equation

$$\frac{CLP_{target}}{CLP_{bsl}} = CLP_{bsd} = e^{-\left[N_0 \cdot \frac{X}{b} \cdot \frac{(1-\rho)^3}{4 \cdot \rho + 1} \right]}$$

to give

$$\frac{X}{b} = -\frac{4 \cdot \rho + 1}{(1 - \rho)^3} \cdot \frac{\ln \left(\dfrac{CLP_{target}}{CLP_{bsl}} \right)}{N_0}$$

Table 12.3 gives the figures for an overall CLP target of 10^{-10}, and Figure 12.7 shows results for three different CLP targets: 10^{-8}, 10^{-10} and 10^{-12}. Figure 12.8 shows results for a range of values of N_0 for an overall CLP target of 10^{-10}.

Table 12.3. Burst-Scale Parameter Values for $N_0 = 100$ and a CLP Target of 10^{-10}

CLP_{bsl}	10^{-1}	10^{-2}	10^{-3}	10^{-4}	10^{-5}	10^{-6}	10^{-7}	10^{-8}	10^{-9}	10^{-10}
load, ρ	0.94	0.87	0.80	0.74	0.69	0.65	0.61	0.58	0.55	0.53
CLP_{bsd}	10^{-9}	10^{-8}	10^{-7}	10^{-6}	10^{-5}	10^{-4}	10^{-3}	10^{-2}	10^{-1}	10^{-0}
X/b	5064.2	394.2	84.6	31.1	14.7	7.7	4.1	2.1	0.8	0

COMBINING THE CONNECTION, BURST AND CELL SCALES

We have seen in Chapter 10 that connection admission control can be based on a variety of different algorithms. An important grade of service parameter, in addition to cell loss and cell delay, is the probability of a connection being blocked. This is very much dependent on the CAC algorithm and the characteristics of the offered traffic types, and in general it is a difficult task to evaluate the connection blocking probability.

However, if we restrict the CAC algorithm to one that is based on limiting the *number* of connections admitted then we can apply erlang's lost call formula to the situation. The service capacity of an ATM link is effectively being divided into N 'circuits'. If all of these 'circuits' are occupied, then the CAC algorithm will reject any further connection attempts. It is worth noting that the cell loss and cell delay performance requirements determine the *maximum* number of connections that can be admitted. Thus, for much of the time, the traffic mix will have fewer connections, and the cell loss and cell delay performance will be rather better than that specified in the traffic contract requirements.

Consider the situation with constant-bit-rate traffic. Figure 12.9 plots the cell loss from a buffer of capacity 10 cells, for a range of CBR sources where D is the number of slots between arrivals. Thus, with a particular CLP requirement, and a constant cell rate given by

$$h = \frac{C}{D} \text{ cell/s}$$

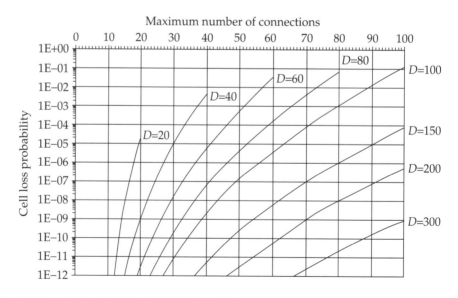

Figure 12.9. Maximum Number of CBR Connections, Given D Cell Slots between Arrivals and CLP

(where C is the cell slot rate), we can find from Figure 12.9 the maximum number of connections that can be admitted onto a link of cell rate C. The link cannot be loaded to more than 100% capacity, so the maximum possible number of sources of any particular type is numerically equal to the (constant) number of cell slots between arrivals. Let's take an example. Suppose we have CBR sources of cell rate 3532 cell/s being multiplexed over a 155.52 Mbit/s link, with a CLP requirement of 10^{-7}. This gives a value of 100 for D, and from Figure 12.9, the maximum number of connections is (near enough) 50.

Now that we have a figure for the maximum number of connections, we can calculate the offered traffic at the connection level for a given probability of blocking. Figure 12.10 shows how the connection blocking probability varies with the maximum number of connections for different offered traffic intensities. With our example, we find that for 50 connections maximum and a connection blocking probability of 0.02, the offered traffic intensity is 40 erlangs. Note that the mean number of connections in progress is numerically equal to the offered traffic, i.e. 40 connections. The cell loss probability for this number of connections can be found from Figure 12.9: it is 2×10^{-9}. This is over an order of magnitude lower than the CLP requirement in the traffic contract, and therefore provides a useful safety margin.

For variable-bit-rate traffic, we will only consider rate envelope multiplexing and not rate sharing. Figure 12.11 shows how the cell loss

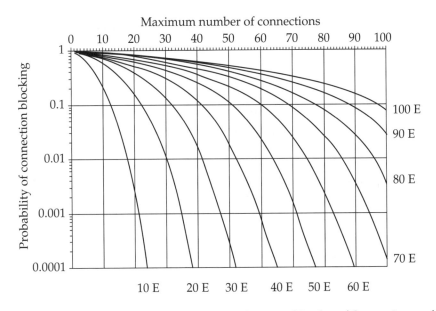

Figure 12.10. Probability of Blocking, Given Maximum Number of Connections and Offered Traffic

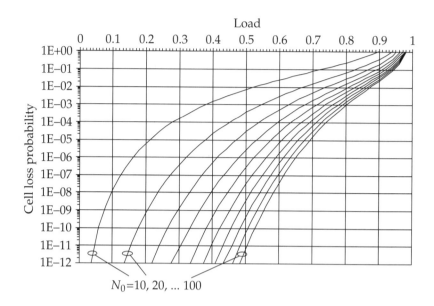

Figure 12.11. Maximum Utilization for VBR Connections, Given N_0 and CLP

probability from a (short) buffer varies with the utilization for a range of VBR sources of different peak cell rates. The key parameter defining this relationship is N_0, the ratio of the cell slot rate to the peak cell rate. Given N_0 and a CLP requirement, we can read off a value for the utilization. This then needs to be multiplied by the ratio of the cell slot rate, C, to the mean cell rate, m, to obtain the maximum number of connections that can be admitted onto the link.

So, for example, for sources with a peak cell rate of 8830 cell/s and a mean cell rate of 3532 cell/s being multiplexed onto a 155.52 Mbit/s link, N_0 is 40 and, according to Figure 12.11, the utilization is about 0.4 for a CLP of 10^{-8}. This utilization is multiplied by 100 (i.e. C/m) to give a maximum of 40 connections. From Figure 12.10, an offered traffic intensity of 30 erlangs gives a connection blocking probability of just under 0.02 for 40 connections maximum. Thus the mean number of connections in progress is 30, giving a mean load of 0.3, and from Figure 12.11 the corresponding cell loss probability is found to be 10^{-11}. This is three orders of magnitude better than for the maximum number of connections.

As an alternative to using Figure 12.10, a traffic table based on Erlang's lost call formula is provided in Table 12.4.

So, we have seen that, for both CBR and VBR traffic, when the connection blocking probability requirements are taken into account, the actual cell loss probability can be rather lower than that for the maximum allowed number of connections.

Table 12.4. Traffic Table Based on Erlang's Lost Call Formula

	Probability of blocking a connection						
N	0.1	0.05	0.02	0.01	0.005	0.001	0.0001
offered traffic in erlangs:							
1	0.11	0.05	0.02	0.01	0.005	0.001	0.0001
2	0.59	0.38	0.22	0.15	0.10	0.04	0.03
3	1.27	0.89	0.60	0.45	0.34	0.19	0.08
4	2.04	1.52	1.09	0.86	0.70	0.43	0.23
5	2.88	2.21	1.65	1.36	1.13	0.76	0.45
6	3.75	2.96	2.27	1.90	1.62	1.14	0.72
7	4.66	3.73	2.93	2.50	2.15	1.57	1.05
8	5.59	4.54	3.62	3.12	2.72	2.05	1.42
9	6.54	5.37	4.34	3.78	3.33	2.55	1.82
10	7.51	6.21	5.08	4.46	3.96	3.09	2.26
11	8.48	7.07	5.84	5.15	4.61	3.65	2.72
12	9.47	7.95	6.61	5.87	5.27	4.23	3.20
13	10.46	8.83	7.40	6.60	5.96	4.83	3.71
14	11.47	9.72	8.20	7.35	6.66	5.44	4.23
15	12.48	10.63	9.00	8.10	7.37	6.07	4.78
16	13.50	11.54	9.82	8.87	8.09	6.72	5.33
17	14.52	12.46	10.65	9.65	8.83	7.37	5.91
18	15.54	13.38	11.49	10.43	9.57	8.04	6.49
19	16.57	14.31	12.33	11.23	10.33	8.72	7.09
20	17.61	15.24	13.18	12.03	11.09	9.41	7.70
21	18.65	16.18	14.03	12.83	11.85	10.10	8.31
22	19.69	17.13	14.89	13.65	12.63	10.81	8.94
23	20.73	18.07	15.76	14.47	13.41	11.52	9.58
24	21.78	19.03	16.63	15.29	14.20	12.24	10.22
25	22.83	19.98	17.50	16.12	14.99	12.96	10.88
26	23.88	20.94	18.38	16.95	15.79	13.70	11.53
27	24.93	21.90	19.26	17.79	16.59	14.43	12.20
28	25.99	22.86	20.15	18.64	17.40	15.18	12.88
29	27.05	23.83	21.03	19.48	18.21	15.93	13.56
30	28.11	24.80	21.93	20.33	19.03	16.68	14.24
31	29.17	25.77	22.82	21.19	19.85	17.44	14.93
32	30.23	26.74	23.72	22.04	20.67	18.20	15.63
33	31.30	27.72	24.62	22.90	21.50	18.97	16.33
34	32.36	28.69	25.52	23.77	22.33	19.74	17.04
35	33.43	29.67	26.43	24.63	23.16	20.51	17.75
36	34.50	30.65	27.34	25.50	24.00	21.29	18.46
37	35.57	31.63	28.25	26.37	24.84	22.07	19.18
38	36.64	32.62	29.16	27.25	25.68	22.86	19.91
39	37.71	33.60	30.08	28.12	26.53	23.65	20.63
40	38.78	34.59	30.99	29.00	27.38	24.44	21.37

(continued overleaf)

Table 12.4. *(continued)*

N	\multicolumn{7}{c}{Probability of blocking a connection}						
	0.1	0.05	0.02	0.01	0.005	0.001	0.0001
	offered traffic in erlangs:						
41	39.86	35.58	31.91	29.88	28.23	25.23	22.10
42	40.93	36.57	32.83	30.77	29.08	26.03	22.84
43	42.01	37.56	33.75	31.65	29.93	26.83	23.58
44	43.08	38.55	34.68	32.54	30.79	27.64	24.33
45	44.16	39.55	35.60	33.43	31.65	28.44	25.08
46	45.24	40.54	36.53	34.32	32.51	29.25	25.83
47	46.32	41.54	37.46	35.21	33.38	30.06	26.58
48	47.40	42.53	38.39	36.10	34.24	30.87	27.34
49	48.48	43.53	39.32	37.00	35.11	31.69	28.10
50	49.56	44.53	40.25	37.90	35.98	32.51	28.86
55	54.9	49.5	44.9	42.4	40.3	36.6	32.7
60	60.4	54.5	49.6	46.9	44.7	40.7	36.6
65	65.8	59.6	54.3	51.5	49.1	44.9	40.5
70	71.2	64.6	59.1	56.1	53.6	49.2	44.5
75	76.7	69.7	63.9	60.7	58.1	53.5	48.6
80	82.2	74.8	68.6	65.3	62.6	57.8	52.6
85	87.6	79.9	73.4	70.0	67.2	62.1	56.7
90	93.1	85.0	78.3	74.6	71.7	66.4	60.9
95	98.6	90.1	83.1	79.3	76.3	70.8	65.0
100	104.1	95.2	87.9	84.0	80.9	75.2	69.2
105	109.5	100.3	92.8	88.7	85.5	79.6	73.4
110	115.0	105.4	97.6	93.4	90.1	84.0	77.6
115	120.5	110.6	102.5	98.2	94.7	88.5	81.9
120	126.0	115.7	107.4	102.9	99.3	92.9	86.2
125	131.5	120.9	112.3	107.7	104.0	97.4	90.4
130	137.0	126.0	117.1	112.4	108.6	101.9	94.7
135	142.5	131.2	122.0	117.2	113.3	106.4	99.0
140	148.1	136.3	126.9	122.0	118.0	110.9	103.4
145	153.6	141.5	131.8	126.7	122.7	115.4	107.7
150	159.1	146.7	136.8	131.5	127.3	119.9	112.1
155	164.6	151.8	141.7	136.3	132.0	124.4	116.4
160	170.1	157.0	146.6	141.1	136.7	129.0	120.8
165	175.6	162.2	151.5	145.9	141.5	133.5	125.2
170	181.1	167.3	156.4	150.7	146.2	138.1	129.6
175	186.7	172.5	161.4	155.5	150.9	142.6	134.0
180	192.2	177.7	166.3	160.4	155.6	147.2	138.4
185	197.7	182.9	171.3	165.2	160.4	151.8	142.8
190	203.2	188.1	176.2	170.0	165.1	156.4	147.2
195	208.7	193.3	181.2	174.9	169.8	161.0	151.7
200	214.0	198.5	186.1	179.7	174.6	165.6	156.1

13 **Priority Control**

the customer comes first

PRIORITIES

ATM networks can feature two forms of priority mechanism: space and time. Both forms relate to how an ATM buffer operates, and these are illustrated in Figure 13.1. Space priority addresses whether or not a cell is admitted into the finite waiting area of the buffer. Time priority deals with the order in which cells leave the waiting area and enter the server for onward transmission. Thus the main focus for the space priority mechanism is to distinguish different levels of cell loss performance, whereas for time priority the focus is on the delay performance. For both forms of priority, the waiting area can be organized in different ways, depending on the specific priority algorithm being implemented.

The ATM standards explicitly support space priority, by the provision of a cell loss priority bit in the ATM cell header. High priority is indicated by the cell loss priority bit having a value of 0, low priority with a value of 1. Different levels of time priority, however, are not explicitly supported in the standards. One way they can be organized is by assigning different levels of time priority to particular VPI/VCI values or ranges of values.

SPACE PRIORITY AND THE CELL LOSS PRIORITY BIT

An ATM terminal distinguishes the level of space priority for the traffic flows it is generating by setting the value of the cell loss priority bit. Within the network, if buffer overflow occurs, the network elements may selectively discard cells of the lower-priority flow in order to maintain the performance objectives required of both the high- and low-priority traffic. For example, a terminal producing compressed video can use high priority for the important synchronization information. This then avoids the need to operate the network elements, through which the video

Space priority
mechanism controls
access to buffer
capacity

Time priority
mechanism controls
access to server
capacity

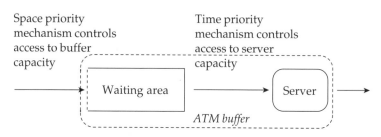

Figure 13.1. Space and Time Priority Mechanisms

connection is routed, at extremely low levels of cell loss probability for all the cells in the connection. The priority mechanism is able to achieve a very low loss probability just for those cells that require it, and this leads to a significant improvement in the traffic load that can be admitted to the network.

Two selective cell discarding schemes have been proposed and studied for ATM buffers: the push-out scheme and partial buffer sharing [13.1]. The push-out scheme is illustrated in Figure 13.2; an arriving cell of high priority which finds the buffer full *replaces* a low-priority cell within the buffer. If the buffer contains only high-priority cells, then the arriving cell is discarded. A low-priority cell arriving to find a full buffer is always discarded. The partial buffer sharing scheme (see Figure 13.3), reserves a part of the buffer for high-priority cells only. If the queue is below a threshold size, then both low- and high-priority cells are accepted onto the queue. Above the threshold only high-priority cells are accepted.

The push-out scheme achieves only slightly better performance than partial buffer sharing. But the buffer management and implementation

The buffer is full with a mix of high
and low priority cells and another
high priority cell arrives

The last low priority cell is 'pushed
out' of the buffer, providing room
for the arriving high priority cell

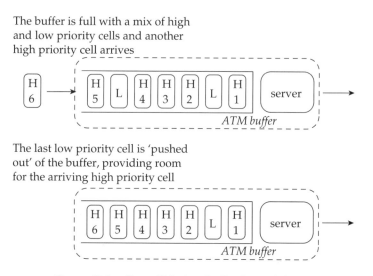

Figure 13.2. Space Priority: the Push-out Scheme

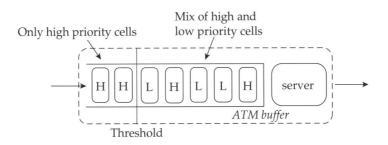

Figure 13.3.   Space Priority: Partial Buffer Sharing

are rather more complex for the push-out mechanism because, when a high-priority cell arrives at a full buffer, a low-priority cell in the buffer must be found and discarded. Thus the partial buffer sharing scheme achieves the best compromise between performance and complexity. Let's look at how partial buffer sharing can be analysed, so we can quantify the improvements in admissible load that are possible with space priorities.

PARTIAL BUFFER SHARING

An analysis of the partial buffer sharing scheme is possible for the sort of queueing system in Chapter 7: a synchronized server, a finite buffer and Poisson input (a synchronized M/D/1/X queue). Here, we will use the line crossing form of analysis (see Chapter 9) as this allows a relatively simple approach.

In Chapter 7, the input traffic is a batch arrival process, where the size of a batch can vary from cell slot to cell slot, described by a probability distribution for the number of cells in the batch. This allows the queue to be analysed for arbitrary distributions, and in Chapter 7 results are shown for Poisson and binomial distributions.

For the analysis of an ATM buffer with partial buffer sharing, we restrict the input to be a Poisson-distributed batch, comprising two streams of traffic: one for each level of space priority. We define the probability that there are k arrivals in one slot as

$$a(k) = \frac{a^k}{k!} \cdot e^{-a}$$

where the mean arrival rate (in cells per cell slot) is given by parameter a. This mean arrival rate is the sum of mean arrival rates, a_h and a_l, for the high- and low-priority streams respectively:

$$a = a_h + a_l$$

and so we can define the probability that there are k high-priority arrivals in one slot as

$$a_h(k) = \frac{a_h^k}{k!} \cdot e^{-a_h}$$

and the probability that there are k low-priority arrivals in one slot as

$$a_l(k) = \frac{a_l^k}{k!} \cdot e^{-a_l}$$

The probability of the queueing system being in state k is defined as

$$s(k) = \Pr\{\text{there are } k \text{ cells, of either priority, in the system}$$

$$\text{at the end of a slot}\}$$

The maximum number of cells in the system, i.e. the waiting area and the server, is X, and the maximum number of low-priority cells, i.e. the threshold level, is M, where $M < X$. Below the threshold level, cells of either priority are admitted into the buffer.

Equating the probabilities of crossing the line between states 0 and 1 gives

$$s(1) \cdot a(0) = s(0) \cdot (1 - a(0))$$

where the left-hand side gives the probability of crossing down (one cell in the queue, which is served, and no arrivals), and the right-hand side gives the probability of crossing up (no cells in the queue, and one or more cells arrive). Remember that any arrivals during the current time slot cannot be served during this slot. Rearranging the equation gives

$$s(1) = \frac{s(0) \cdot (1 - a(0))}{a(0)}$$

In general, equating the probabilities of crossing the line between states $k - 1$ and k, for $k < M$, gives

$$s(k) \cdot a(0) = s(0) \cdot A(k) + \sum_{i=1}^{k-1} s(i) \cdot A(k - i + 1)$$

where $A(k)$ is the probability that at least k cells arrive during the time slot, and is expressed simply as the probability that fewer than k cells do not arrive.

$$A(k) = 1 - \sum_{j=0}^{k-1} a(j)$$

$A_h(k)$ is the probability that at least k high-priority cells arrive during a time slot, and is defined in a similar manner in terms of $a_h(j)$; this is used later on in the analysis.

So, in general for $k < M$, we have

$$s(k) = \frac{s(0) \cdot A(k) + \sum_{i=1}^{k-1} s(i) \cdot A(k - i + 1)}{a(0)}$$

Continuing the analysis for state probabilities $s(k)$ at or above $k = M$ is not so straightforward, because the order in which the cells arrive in the buffer is important if the system is changing from a state below the threshold to a state above the threshold.

Consider the case in which a buffer, with a threshold $M = 10$ cells and system capacity $X = 20$ cells, has 8 cells in the system at the end of a time slot. During the next time slot, 4 low-priority cells and 2 high-priority cells arrive, and one cell is served. If the low-priority cells arrive first, then 2 low-priority cells are admitted, taking the system up to the threshold, the other 2 low-priority cells are discarded, and the 2 high-priority cells are admitted, taking the system size to 12. Then the cell in the server completes service and the system size reduces to 11, which is the system state at the end of this time slot. If the high-priority cells arrive first, then these take the system up to the threshold size of 10, and so all 4 low-priority cells are discarded. At the end of the slot the system size is then 9 (the cell in the server completes service).

To analyse how the system changes from one state to another we need to know the number of cells that are *admitted* onto the buffer (at a later stage we will be interested in the number of cells that are *not* admitted, in order to calculate the loss from the system). So, let's say that $m + n$ cells are admitted out of a total of i cells that arrive during one cell slot. Of those admitted, the first m are of either high or low priority and take the system from its current state up to the threshold level, and then the other n are of high priority. Thus $i - (m + n)$ low-priority cells are lost. We use the following expression for the probability that these $m + n$ cells are admitted:

$$a'(m, n) = \sum_{i=m+n}^{\infty} \left[a(i) \cdot \frac{(i - m)!}{n! \cdot (i - m - n)!} \cdot \left(\frac{a_h}{a}\right)^n \cdot \left(\frac{a_l}{a}\right)^{i-m-n} \right]$$

The binomial part of the expression determines the probability that, of the $i - m$ cells to arrive when the queue is at or above the threshold, n are high-priority cells. Here, the probability that a cell is of high priority is expressed as the proportion of the mean arrival rate that is of high priority. Note that although this expression is an infinite summation, it converges rapidly and so needs only a few terms to obtain a value for $a'(m, n)$.

With the line crossing analysis, we need to express the probability that m cells of either priority arrive, and then at least n or more high-priority cells arrive, denoted $A'(m, n)$. This can be expressed as

$$A'(m, n) = \sum_{j=n}^{\infty} a'(m, j)$$

Another way of expressing this is by working out the probability that fewer than $m + n$ cells are admitted. This happens in two different ways: either the total number of cells arriving during a slot is not enough, or there are enough cells but the order in which they arrive is such that there are not enough high-priority cells above the threshold.

$$A'(m, n) = 1 - \sum_{i=0}^{m+n-1} a(i) - \sum_{i=0}^{\infty} \left[a(m + n + i) \cdot \sum_{j=0}^{n-1} \left\{ \frac{(n+i)!}{j! \cdot (n+i-j)!} \right. \right.$$
$$\left. \left. \cdot \left(\frac{a_h}{a} \right)^j \cdot \left(\frac{a_l}{a} \right)^{n+i-j} \right\} \right]$$

We can now analyse the system at or above the threshold. Equating probabilities of crossing the line between M and $M - 1$ gives

$$s(M) \cdot a_h(0) = s(0) \cdot A(M) + \sum_{i=1}^{M-1} s(i) \cdot A'(M - i, 1)$$

The left-hand side is the probability of crossing down; to stay at the threshold level, or to move above it, at least one high-priority cell is needed, so the state reduces by one only if there are no high-priority arrivals. The right-hand side is the probability of crossing up from below the threshold. The first term is for crossing up from the state when there is nothing in the system; this requires M, or more, cells of either priority. The second term is for all the non-zero states, i, below the threshold; in these cases there is always a cell in the server which leaves the system after any arrivals have been admitted to the queue. Thus at least one high-priority arrival is required after there have been sufficient arrivals ($M - i$) of either priority to fill the queue up to the threshold.

For $k > M$, we have

$$s(k) \cdot a_h(0) = s(0) \cdot A'(M, k - M) + \sum_{i=1}^{M-1} \{s(i) \cdot A'(M - i, k - M + 1)\}$$

$$+ \sum_{i=M}^{k-1} \{s(i) \cdot A_h(k - i + 1)\}$$

This differs from the situation for $k = M$ in two respects: first, the crossing up from state 0 requires M cells of either priority and a further $k - M$ of high-priority; and secondly, it is now possible to cross the line from a state at or above the threshold – this can only be achieved with high-priority arrivals.

At the buffer limit, $k = X$, we have only one way of reaching this state: from state 0, with M cells of either priority followed by at least $X - M$ cells of high-priority. If there is at least one cell in the queue at the start of the slot, and enough arrivals fill the queue, then at the end of the slot, the cell in the server will complete service and take the queue state from X down to $X - 1$. Thus for $k = X$ we have

$$s(X) \cdot a_h(0) = s(0) \cdot A'(M, X - M)$$

Now, as in Chapter 7, we have no value for $s(0)$, so we cannot evaluate $s(k)$ for $k > 0$. Therefore we define a new variable, $u(k)$, as

$$u(k) = \frac{s(k)}{s(0)}$$

so

$$u(0) = 1$$

Then

$$u(1) = \frac{(1 - a(0))}{a(0)}$$

For $1 < k < M$

$$u(k) = \frac{A(k) + \sum_{i=1}^{k-1} u(i) \cdot A(k - i + 1)}{a(0)}$$

At the threshold

$$u(M) = \frac{A(M) + \sum_{i=1}^{M-1} u(i) \cdot A'(M - i, 1)}{a_h(0)}$$

For $M < k < X - 1$

$$u(k) = \frac{A'(M, k - M) + \sum_{i=1}^{M-1} \{u(i) \cdot A'(M - i, k - M + 1)\} + \sum_{i=M}^{k-1} \{u(i) \cdot A_h(k - i + 1)\}}{a_h(0)}$$

At the system capacity

$$u(X) = \frac{A'(M, X - M)}{a_h(0)}$$

All the values of $u(k)$, $0 \leqslant k \leqslant X$, can be evaluated. Then, as in Chapter 7, we can calculate the probability that the system is empty:

$$s(0) = \frac{1}{\sum_{i=0}^{X} u(i)}$$

and, from that, find the rest of the state probability distribution:

$$s(k) = s(0) \cdot u(k)$$

Before we go on to calculate the cell loss probability for the high-and low-priority cell streams, let's first show an example state probability distribution for an ATM buffer implementing the partial buffer sharing scheme. Figure 13.4 shows the state probabilities when the buffer capacity is 20 cells, and the threshold level is 15 cells, for three different loads: (i) the low priority load, a_l, is 0.7 and the high-priority load, a_h, is 0.175 of the cell slot rate; (ii) $a_l = 0.6$ and $a_h = 0.15$; and (iii) $a_l = 0.5$ and $a_h = 0.125$.

The graph shows a clear distinction between the gradients of the state probability distribution below and above the threshold level. Below the threshold, the queue behaves like an ordinary M/D/1 with a gradient corresponding to the combined high- and low-priority load. Above the threshold, only the high-priority cell stream has any effect, and so the gradient is much steeper because the load on this part of the queue is much less.

In Chapter 7, the loss probability was found by comparing the offered and the carried traffic at the cell level. But now we have two different priority streams, and the partial buffer sharing analysis only gives the combined carried traffic. The *overall* cell loss probability can be found

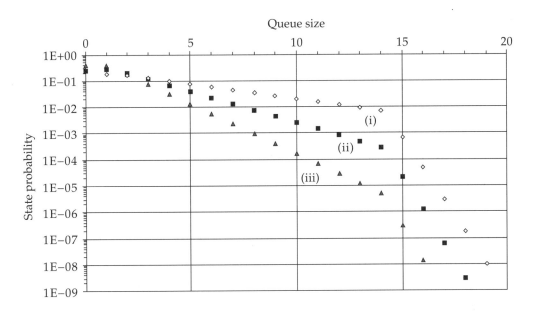

Figure 13.4. State Probability Distribution for ATM Buffer with Partial Buffer Sharing (i) $a_l = 0.7$, $a_h = 0.175$; (ii) $a_l = 0.6$, $a_h = 0.15$; (iii) $a_l = 0.5$, $a_h = 0.125$

using

$$\text{CLP} = \frac{a_l + a_h - (1 - s(0))}{a_l + a_h}$$

But the main objective of having a space priority scheme is to provide different levels of cell loss. How can we calculate this cell loss probability for each priority stream? It has to be done by considering the probability of losing a group of low- or high-priority cells during a cell slot, and then taking the weighted mean over all the possible group sizes. The high-priority cell loss probability is given by

$$\text{CLP}_h = \frac{\sum_j j \cdot l_h(j)}{a_h}$$

where $l_h(j)$ is the probability that j high-priority cells are lost in a cell slot and is given by

$$l_h(j) = \sum_{i=0}^{M-1} s(i) \cdot a'(M - i, X - M + j) + \sum_{i=M}^{X} s(i) \cdot a_h(X - i + j)$$

The first summation on the right-hand side accounts for the different ways of losing j cells when the state of the system is less than the threshold.

This involves filling up to the threshold with either low- or high-priority cells, followed by $X - M$ high-priority cells to fill the queue and then a further j high-priority cells which are lost. The second summation deals with the different ways of losing j cells when the state of the system is at or above the threshold; $X - i$ high-priority cells are needed to fill the queue and the other j in the batch are lost.

The low-priority loss is found in a similar way:

$$CLP_l = \frac{\sum_j j \cdot l_l(j)}{a_l}$$

where $l_l(j)$ is the probability that j low-priority cells are lost in a cell slot and is given by

$$l_l(j) = \sum_{i=0}^{M-1} \left[s(i) \cdot \sum_{r=M-i+j}^{\infty} a(r) \cdot \frac{(r-(M-i))!}{(r-(M-i)-j)! \cdot j!} \cdot \left(\frac{a_h}{a}\right)^{r-(M-i)-j} \cdot \left(\frac{a_l}{a}\right)^j \right]$$
$$+ \sum_{i=M}^{X} s(i) \cdot a_l(j)$$

The first term on the right-hand side accounts for the different ways of losing j cells when the state of the system is less than the threshold. This involves filling up to the threshold with either $M - i$ cells of either low or high-priority, followed by any number of high-priority cells along with j low-priority cells (which are lost). The second summation deals with the different ways of losing j cells when the state of the system is above the threshold. This is simply the probability of j low-priority cells arriving in a time slot, for each of the states at or above the threshold.

Increasing the admissible load

Let's now demonstrate the effect of introducing a partial buffer sharing mechanism to an ATM buffer. Suppose we have a buffer of size $X = 20$, and the most stringent cell loss probability requirement for traffic through the buffer is 10^{-10}. From Table 10.1 we find that the maximum admissible load is 0.521. Now the traffic mix is such that there is a high-priority load of 0.125 which requires the CLP of 10^{-10}; the rest of the traffic can tolerate a CLP of 10^{-3}, a margin of seven orders of magnitude. Without a space priority mechanism, a maximum load of $0.521 - 0.125 = 0.396$ of this other traffic can be admitted. However, the partial buffer sharing analysis shows that, with a threshold of $M = 15$, the low-priority load can be increased to 0.7 to give a cell loss probability of 1.16×10^{-3}, and the high-priority load of 0.125 has a cell loss probability of 9.36×10^{-11}. The total admissible load has increased by just over 30% of the cell slot rate, from 0.521 to 0.825, representing a 75% increase in the low-priority traffic.

If the threshold is set to $M = 18$, the low-priority load can only be increased to 0.475 giving a cell loss probability of 5.6×10^{-8}, and the high-priority load of 0.125 has a cell loss probability of 8.8×10^{-11}. But even this is an extra 8% of the cell slot rate, representing an increase in 20% for the low-priority traffic, for a cell loss margin of between two and three orders of magnitude. Thus a substantial increase in load is possible, particularly if the difference in cell loss probability requirement is large.

Dimensioning buffers for partial buffer sharing

Figures 13.5 and 13.6 show interesting results from the partial buffer sharing analysis. In both cases, the high-priority load is fixed at 0.125, and the space above the threshold is held constant at 5 cells. In Figure 13.5, the low-priority load is varied from 0.4 up to 0.8, and the cell loss probability results are plotted for the high- and low-priority traffic against the combined load. This is done for three different buffer capacities. The results show that the margin in the cell loss probabilities is almost constant, at seven orders of magnitude. Figure 13.6 shows the same margin in the cell loss probabilities for a total load of 0.925 ($a_h = 0.125$, $a_l = 0.8$) as the buffer capacity is varied from 10 cells up to 50 cells.

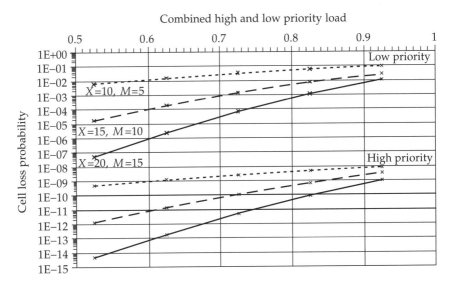

Figure 13.5. Low and High-Priority Cell Loss against Load, for $X - M = 5$ and $a_h = 0.125$

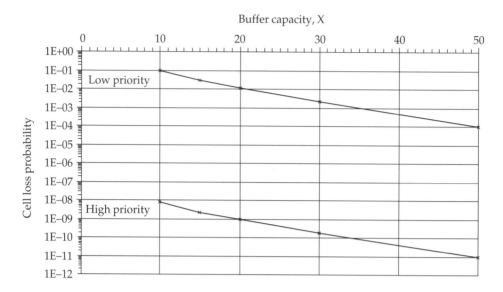

Figure 13.6. Low- and High-Priority Cell Loss against Buffer Capacity, for $a = 0.925$ and $X - M = 5$

The difference between high- and low-priority cell loss is almost invariant to the buffer capacity and the total load, provided that the space above the threshold, and the high-priority load, are kept constant. Table 13.1 shows how the margin varies with the space above the threshold, and the high-priority load (note that margins greater than 11 orders of magnitude are not included – these are unlikely to be required in practice). The values are also plotted in Figure 13.7.

With this information, buffers can be dimensioned using the following procedure:

1. Set the threshold by using Table 10.1 based on the $M/D/1/X$ analysis (without priorities) for the combined load and the combined cell loss probability requirement. The latter is found using the following relationship (which is based on equating the average number of cells

Table 13.1. Cell Loss Probability Margin between Low- and High-Priority Traffic

$X - M$	High-priority traffic load (as a fraction of the cell slot rate)								
	0.01	0.02	0.03	0.04	0.05	0.1	0.15	0.2	0.25
1	2.7E-03	5.6E-03	8.4E-03	1.1E-02	1.4E-02	2.8E-02	4.4E-02	5.9E-02	7.6E-02
2	6.5E-06	2.7E-05	6.3E-05	1.2E-04	1.8E-04	8.4E-04	2.2E-03	4.3E-03	7.5E-03
3	1.4E-08	1.2E-07	4.5E-07	1.2E-06	2.5E-06	2.6E-05	1.1E-04	3.2E-04	7.4E-04
4	3.0E-11	5.4E-10	2.9E-09	1.0E-08	2.8E-08	6.7E-07	4.9E-06	2.2E-05	7.0E-05
5			1.8E-11	9.0E-11	3.3E-10	1.9E-08	2.5E-07	1.6E-06	7.0E-06

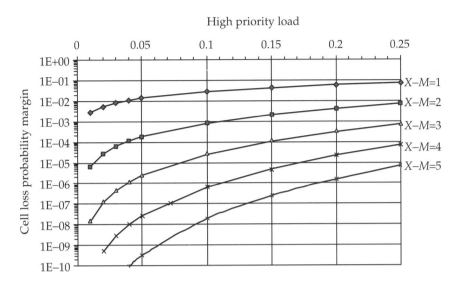

High priority load

Figure 13.7. Cell Loss Probability Margin against High-Priority Load for Different Values of $X - M$

lost per cell slot):

$$CLP = \frac{a_l \cdot CLP_l + a_h \cdot CLP_h}{a}$$

2. Add buffer space above the threshold determined by the high-priority load and the additional cell loss probability margin, from Table 13.1.

Let's take an example. We have a requirement for a buffer to carry a total load of 0.7, with low-priority CLP of 10^{-6} and high-priority CLP of 10^{-10}. The high-priority load is 0.15. Thus the overall CLP is given by

$$CLP = \frac{0.55 \times 10^{-6} + 0.15 \times 10^{-10}}{0.7} = 7.86 \times 10^{-7}$$

From Table 10.1 we find that the threshold is between 20 and 25 cells, but closer to 20; we will use $M = 21$. Table 13.1 gives an additional buffer space of 3 cells for a margin of 10^{-4} and high-priority load of 0.15. Thus the total buffer capacity is 24. If we put these values of $X = 24$ and $M = 21$, $a_l = 0.55$ and $a_h = 0.15$ back into the analysis, the results are $CLP_l = 5.5 \times 10^{-7}$ and $CLP_h = 6.3 \times 10^{-11}$. For a buffer size of 23, a threshold of 20, and the same load, the results are $CLP_l = 1.1 \times 10^{-6}$ and $CLP_h = 1.2 \times 10^{-10}$.

Two of the values for high-priority load in Table 13.1 are of particular interest in the development of a useful dimensioning rule; these values are 0.04 and 0.25. In Figure 13.8, the CLP margin is plotted against the

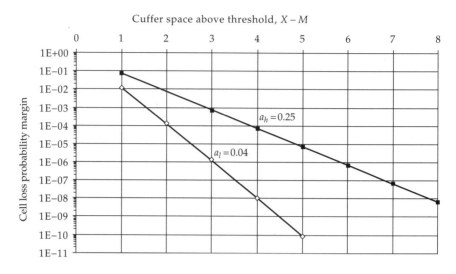

Figure 13.8. Cell Loss Probability Margin against Buffer Space Reserved for High-Priority Traffic, $X - M$

buffer space above the threshold (this is shown as a continuous line to illustrate the log-linear relationship–the buffer space of course varies in integer values). At the 25% load, each cell space reserved for high-priority traffic is worth one order of magnitude on the CLP margin. At the 4% load, it is two orders of magnitude. We can express this as

$$CLP_{margin} = 10^{-(X-M)}$$

for a high-priority load of 25% of the cell slot rate, and

$$CLP_{margin} = 10^{-2 \cdot (X-M)}$$

for a high-priority load of 4% of the cell slot rate.

TIME PRIORITY IN ATM

In order to demonstrate the operation of time priorities, let's define two traffic classes, of high and low time priority. In a practical system, there may be rather more levels, according to the perceived traffic requirements. The ATM buffer in Figure 13.9 operates in such a way that any high-priority cells are always served before any low-priority. Thus a high-priority cell arriving at a buffer with only low-priority cells currently in the queue will go straight to the head of the queue. Note that at the beginning of time slot $n + 1$ the low-priority cell currently at the head of

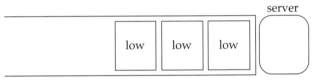

State of the buffer at end of time slot n

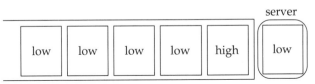

State of the buffer at end of time slot $n+1$ after 3 cells
have arrived - one of high priority and two of low priority

Figure 13.9. Time Priorities in ATM

the queue goes into service. It is only during time slot $n + 1$ that the high-priority cell arrives and is then placed at the head of the queue. The same principle can be applied with many levels of priority. Note that any cell arriving to find the buffer full is lost, regardless of the level of time priority.

The effect of time priorities is to decrease the delay for the higher-priority traffic at the expense of increasing the delays for the lower-priority traffic. As far as ATM is concerned, this means that real-time connections (e.g. voice and interactive video) can be speeded on their way at the expense of delaying the cells of connections which do not have real-time constraints, e.g. email data.

To analyse the delay performance for a system with two levels of time priority, we will assume an M/D/1 system, with infinite buffer length. Although time priorities do affect the cell loss performance, we will concentrate on those analytical results that apply to *delay*.

Mean value analysis

We define the mean arrival rate in cells per slot as a_i for cells of priority i. High-priority is indicated by $i = 1$ and low priority by $i = 2$. Note that the following analysis can be extended to many levels if required.

The formulas for the mean waiting time are:

$$w_1 = \frac{a_1 + a_2}{2 \cdot (1 - a_1)}$$

and

$$w_2 = \frac{w_1 \cdot a_1 + \dfrac{a_1 + a_2}{2}}{1 - a_1 - a_2}$$

where w_i is the mean wait (in time slots) endured by cells of priority i while in the buffer.

Consider an ATM scenario in which a very small proportion of traffic, say about 1%, is given high time priority. Figure 13.10 shows the effect on the mean waiting times. Granting a time priority to a small proportion of traffic has very little effect on the mean wait for the lower-priority traffic, which is indistinguishable from the mean wait when there are no priorities. We can also see from the results that the waiting time for the high-priority cells is greatly improved.

Figure 13.11 shows what happens if the proportion of high-priority traffic is significantly increased, to 50% of the combined high- and low-priority load. Even in this situation, mean waiting times for the low-priority cells are not severely affected, and waiting times for the priority traffic have still been noticeably improved. Figure 13.12 illustrates the case when most of the traffic is high-priority and only 1% is of low priority. Here, there is little difference between the no-priority case and the results for the high-priority traffic, but the very small amount of low-priority traffic has significantly worse waiting times.

The results so far are for the *mean* waiting time. Let's now consider the effect of time priorities on the *distribution* of the waiting-time (see also [13.2]). To find the waiting time probabilities for cells in an ATM

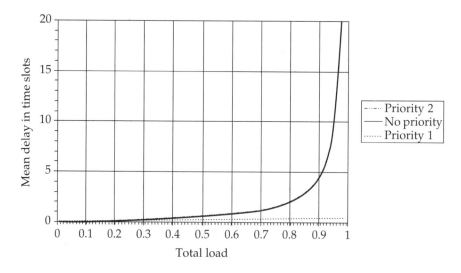

Figure 13.10. Mean Waiting Time for High and Low Time-Priority Traffic, where the Proportion of High-Priority Traffic is 1% of the Total Load

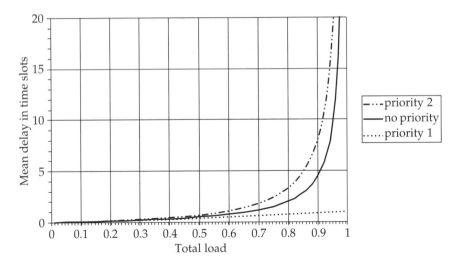

Figure 13.11. Mean Waiting Time for High and Low Time-Priority Traffic, where the Proportion of High-Priority Traffic is 50% of the Total Load

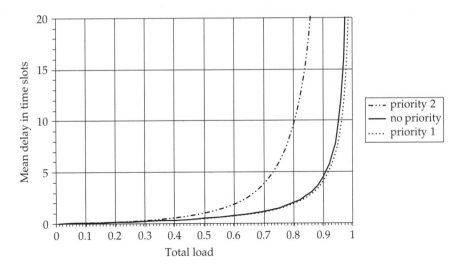

Figure 13.12. Mean Waiting Time for High and Low Time-Priority Traffic, where the Proportion of High-Priority Traffic is 99% of the Total Load

buffer where different levels of time priority are present requires the use of convolution (as indeed did finding waiting times in a non-priority buffer–see Chapter 7). A cell, say **C**, arriving in time slot i will wait behind a number of cells, and this number has four components:

1. the total number of cells of equal or higher priority that are present in the buffer at the end of time slot $i - 1$

2. the number of cells of the *same priority* that are ahead of **C** in the batch in which **C** arrives

3. all the cells of higher priority than **C** that arrive in time slot i

4. higher-priority cells that arrive subsequently, but before **C** enters service.

Again, if we focus on just two levels of priority, we can find the probability that a cell of low priority (priority 2) has to wait k time slots before it can enter service, by finding expressions for the four individual components. Let us define:

component 1   the unfinished work—as $u(k)$

component 2   the 'batch wait'—as $b(k)$

component 3   the wait caused by priority-1 arrivals in time slot i – as $a_1(k)$

Then, excluding the effect of subsequent high-priority arrivals, we know that our waiting-time distribution must be (in part) the sum of the three components listed above. Note that to sum random variables, we must convolve their distributions. We will call the result of this convolution the 'virtual waiting-time distribution', v(k), given by:

$$v(\cdot) = u(\cdot) * b(\cdot) * a_1(\cdot)$$

where * denotes convolution. We can rewrite this as:

$$v(k) = \sum_{i=0}^{k} \left[u(k - i) \cdot \sum_{j=0}^{i} b(j) \cdot a_1(i - j) \right]$$

But where do the three distributions, $u(k)$, $b(k)$ and $a_1(k)$ come from? As we are assuming Poisson arrivals for both priorities, $a_1(k)$ is simply:

$$a_1(k) = \frac{a_1^k}{k!} \cdot e^{-a_1}$$

and for the low-priority cells we will have:

$$a_2(k) = \frac{a_2^k}{k!} \cdot e^{-a_2}$$

where

a_1 is the arrival rate (in cells per time slot) of high-priority cells

a_2 is the arrival rate (in cells per time slot) of low-priority cells

The unfinished work, $u(k)$, is actually found from the state probabilities, denoted $s(k)$, the formula for which was given in Chapter 7:

$$u(0) = s(0) + s(1)$$

$$u(k) = s(k + 1) \quad \text{for } k > 0$$

What about the wait caused by other cells of low priority arriving in the same batch, but in front of **C**? Well, there is a simple approach here too:

$$b(k) = \Pr\{\mathbf{C} \text{ is } (k + 1)^{\text{th}} \text{ in the batch}\}$$

$$= \frac{E[\text{number of cells that are } (k + 1)^{\text{th}} \text{ in their batch}]}{E[\text{number of cells arriving per slot}]}$$

$$= \frac{1 - \displaystyle\sum_{i=0}^{k} a_2(i)}{a_2}$$

So now all the parts are assembled, and we need only implement the convolution to find the virtual waiting-time distribution. However, this still leaves us with the problem of accounting for subsequent high-priority arrivals. In fact, this is very easy to do using a formula developed (originally) for an entirely different purpose. The result is that:

$$w(0) = v(0)$$

$$w(k) = \frac{\displaystyle\sum_{i=1}^{k} v(i) \cdot a_1(k - i, k) \cdot i}{k} \quad \text{for } k > 0$$

where:

$$w(k) = \Pr\{\text{a priority 2 cell must wait } k \text{ time slots before it enters service}\}$$

$$a_1(k, x) = \frac{a_1^{x \cdot k}}{x!} \cdot e^{-k \cdot a_1}$$

So $a_1(k, x)$ is simply the probability that k priority-1 cells arrive in x time slots.

Figures 13.13 and 13.14 show the waiting-time distributions for high- and low-priority cells when the combined load is 0.8 cells per time slot and the high-priority proportions are 1% and 50% respectively. From these results, it is clear that, even for a relatively large proportion of high-priority traffic, the effect on the waiting-time distribution for low-priority traffic is small, but the benefits to the high-priority traffic are significant.

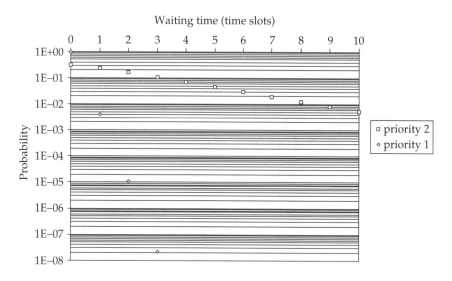

Figure 13.13. Waiting-Time Distribution for High- and Low Time-Priority Traffic, where the Proportion of High-Priority Traffic Is 1% of a Total Load of 0.8 Cells per Time Slot

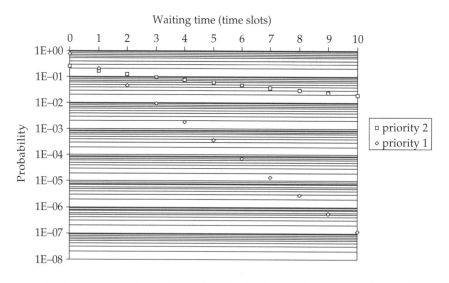

Figure 13.14. Waiting Time Distribution for High and Low Time-Priority Traffic, where the Proportion of High-Priority Traffic Is 50% of a Total Load of 0.8 Cells per Time Slot

Before we leave time priorities, it is worth noting that practical systems for implementing them would probably feature a buffer for each priority level, as shown in Figure 13.15, rather than one buffer for all priorities, as in Figure 13.9. Although there is no explicit provision in the Standards for distinguishing different levels of time-priority, it is possible to use the

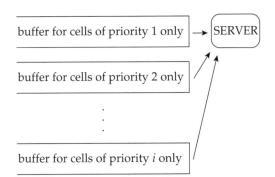

Figure 13.15. Practical Arrangement of Priority Buffers at an Output Port

VPI/VCI values in the header. On entry to a switch, the VPI/VCI values are used to determine the outgoing port required, so it is a relatively simple extension to use these values to choose one of a number of time priority buffers at that output port.

Using one buffer per priority level (Figure 13.15) would have little effect on the delays experienced by the cells but it would affect the CLP. This is because, for a given total capacity of X cells, the CLP is minimized if the total space is shared amongst the different priorities (as in Figure 13.9). However, this has to be balanced against considerations of extra complexity (and hence cost) inherent in a buffer-sharing arrangement.

PART III

IP Performance and Traffic Management

14 Basic Packet Queueing

the long and short of it

THE QUEUEING BEHAVIOUR OF PACKETS IN AN IP ROUTER BUFFER

In Chapters 7 and 8, we investigated the basic queueing behaviour found in ATM output buffers. This queueing arises because multiple streams of cells are being multiplexed together; hence the need for (relatively short) buffers. We developed balance equations for the state of the system at the end of any time slot, from which we derived cell loss and delay results. We also looked at heavy-traffic approximations: explicit equations which could be rearranged to yield expressions for buffer dimensioning and admission control, as well as performance evaluation.

In essence, packet queueing is very similar. An IP router forwards arriving packets from input port to output port: the queueing behaviour arises because multiple streams of packets (from different input ports) are being multiplexed together (over the same output port). However, a key difference is that packets do not all have the same length. The minimum header size in IPv4 is 20 octets, and in IPv6, it is 40 octets; the maximum packet size depends on the specific sub-network technology (e.g. 1500 octets in Ethernet, and 1000 octets is common in X.25 networks). This difference has a direct impact on the service time; to take this into account we need a probabilistic (rather than deterministic) model of service, and a different approach to the queueing analysis.

As before, there are three different types of behaviour in which we are interested:

- the state probabilities, by which we mean the proportion of time that a queue is found to be in a particular state (being in state k means the queue contains k packets at the time at which it is inspected, and measured over a very long period of time, i.e. the *steady-state* probabilities);

- the packet loss probability, by which we mean the proportion of packets lost over a very long period of time;

- the packet waiting-time probabilities, by which we mean the probabilities associated with a packet being delayed k time units.

It turns out that accurate evaluation of the state probabilities is paramount in calculating the waiting times and loss too, and for this reason we focus on finding accurate and simple-to-use formulas for state probabilities.

BALANCE EQUATIONS FOR PACKET BUFFERING: THE GEO/GEO/1

To analyse these different types of behaviour, we are going to start by following the approach developed in Chapter 7, initially for a very simple queue model called the Geo/Geo/1, which is the discrete-time version of the 'classical' queue model M/M/1. One way in which this model differs from that of Chapter 7 is that the fundamental time unit is reduced from a cell service time to the time to transmit an octet (byte), T_{oct}. Thus we have a 'conveyor belt' of octets – the transmission of each octet of a packet is synchronized to the start of transmission of the previous octet. Using this model assumes a geometric distribution as a first attempt at variable packet sizes:

$$b(k) = \Pr\{\text{packet size is } k \text{ octets}\} = (1 - q)^{k-1} \cdot q$$

where

$$q = \Pr\{\text{a packet completes service at the end of an octet slot}\}$$

We use a Bernoulli process for the packet arrivals, i.e. a geometrically distributed number of slots between arrivals (the first Geo in Geo/Geo/1):

$$p = \Pr\{\text{a packet arrives in an octet slot}\}$$

Thus we have an independent and identically distributed batch of k octets ($k = 0, 1, 2, \ldots$) arriving in each octet slot:

$$a(0) = \Pr\{\text{no octets arriving in an octet slot}\} = 1 - p$$

$$a(k) = \Pr\{k > 0 \text{ octets in an octet slot}\} = p \cdot b(k)$$

The mean service time for a packet is simply the mean number of octets (the inverse of the exit probability for the geometric distribution, i.e. $1/q$) multiplied by the octet transmission time.

$$s = \frac{T_{oct}}{q}$$

giving a packet service rate of

$$\mu = \frac{1}{s} = \frac{q}{T_{oct}}$$

The mean arrival rate is

$$\lambda = \frac{p}{T_{oct}}$$

and so the applied load is given by

$$\rho = \frac{\lambda}{\mu} = \frac{p}{q}$$

This is also the utilization, assuming an infinite buffer size and, hence, no packet loss. We define the state probability, i.e. the probability of being in state k, as

$$s(k) = \text{Pr}\{\text{there are } k \text{ octets in the queueing system at the}$$

$$\text{end of any octet slot}\}$$

As before, the utilization is just the steady-state probability that the system is not empty, so

$$\rho = 1 - s(0)$$

and therefore

$$s(0) = 1 - \frac{p}{q}$$

Calculating the state probability distribution

As in Chapter 7, we can build on this value, $s(0)$, by considering all the ways in which it is possible to reach the empty state:

$$s(0) = s(0) \cdot a(0) + s(1) \cdot a(0)$$

giving

$$s(1) = s(0) \cdot \frac{1 - a(0)}{a(0)} = \left(1 - \frac{p}{q}\right) \cdot \frac{p}{1 - p}$$

Similarly, we find a formula for $s(2)$ by writing the balance equation for $s(1)$, and rearranging:

$$s(2) = \frac{s(1) - s(0) \cdot a(1) - s(1) \cdot a(1)}{a(0)}$$

which, after substituting in

$$a(0) = 1 - p$$
$$a(1) = p \cdot q$$

gives

$$s(2) = s(1) \cdot \left(\frac{1-q}{1-p}\right) = \left(1 - \frac{p}{q}\right) \cdot \frac{p}{1-p} \cdot \frac{1-q}{1-p}$$
$$= \left(1 - \frac{p}{q}\right) \cdot \frac{p}{1-q} \cdot \left(\frac{1-q}{1-p}\right)^2$$

By induction, we find that

$$s(k) = \left(1 - \frac{p}{q}\right) \cdot \frac{p}{1-q} \cdot \left(\frac{1-q}{1-p}\right)^k \quad \text{for } k > 0$$

As in Chapter 7, the state probabilities refer to the state of the queue at moments in time that are the 'end of time unit instants'.

We can take the analysis one step further to find an expression for the probability that the queue exceeds k octets, $Q(k)$:

$$Q(k) = 1 - s(0) - s(1) - \cdots - s(k)$$

This gives a geometric progression which, after some rearrangement, yields

$$Q(k) = \frac{p}{q} \cdot \left(\frac{1-q}{1-p}\right)^k$$

To express this in terms of packets, x, (recall that it is currently in terms of octets), we can simply substitute

$$k = x \cdot (\text{mean number of octets per packet}) = x \cdot \frac{1}{q}$$

giving an expression for the probability that the queue exceeds x packets:

$$Q(x) = \frac{p}{q} \cdot \left(\frac{1-q}{1-p}\right)^{\frac{x}{q}}$$

So, what do the results look like? Let's use a load of 80%, for comparison with the results in Chapter 7, and assume an average packet size of

500 octets. Thus

$$\rho = \frac{p}{q} = 0.8$$

$$\frac{1}{q} = 500 \Rightarrow q = 0.002$$

$$p = 0.8 \times 0.002 = 0.0016$$

The results are shown in Figure 14.1, labelled Geo/Geo/1. Those labelled 'Poisson' and 'Binomial' are the results from Chapter 7

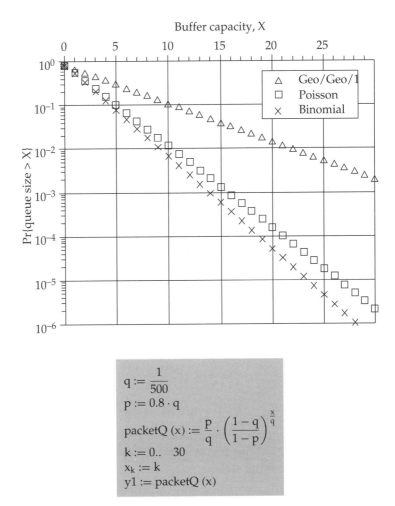

Figure 14.1. Graph of the Probability that the Queue State Exceeds X, and the Mathcad Code to Generate (x, y) Values for Plotting the Geo/Geo/1 Results. For Details of how to Generate the Results for Poisson and Binomial Arrivals to a Deterministic Queue, see Figure 7.6

(Figure 7.6) for fixed service times at a load of 80%. Notice that the variability in the packet sizes (and hence service times) produces a flatter gradient than the fixed-cell-size analysis for the same load. The graph shows that, for a given performance requirement (e.g. 0.01), the buffer needs to be about twice the size ($X = 21$) of that for fixed-size packets or cells ($X = 10$). This corresponds closely with the difference, in average waiting times, between $M/D/1$ and $M/M/1$ queueing systems mentioned in Chapter 4.

DECAY RATE ANALYSIS

One of the most important effects we have seen so far is that the state probability values we are calculating tend to form straight lines when the queue size (state) is plotted on a linear scale, and the state probability is plotted on a logarithmic scale. This is a very common (almost universal) feature of queueing systems, and for this reason has become a key result that we can use to our advantage.

As in the previous section, we define the state probability as

$$s(k) = \Pr\{\text{there are } k \text{ units of data} - \text{packets,}$$

$$\text{octets} - \text{in the queueing system}\}$$

We define the 'decay rate' (DR) as the ratio:

$$\frac{s(k+1)}{s(k)}$$

However, this ratio will not necessarily stay constant until k becomes large enough, so we should actually say that:

$$\text{DR} = \frac{s(k+1)}{s(k)} \quad \text{as } k \to \infty$$

as illustrated in Figure 14.2.

From the form of the equation, and the example parameter values in Figure 14.1, we can see that the decay rate for the Geo/Geo/1 model is constant from the start:

$$\frac{s(k+1)}{s(k)} = \frac{\left(1 - \dfrac{p}{q}\right) \cdot \dfrac{p}{1-q} \cdot \left(\dfrac{1-q}{1-p}\right)^{k+1}}{\left(1 - \dfrac{p}{q}\right) \cdot \dfrac{p}{1-q} \cdot \left(\dfrac{1-q}{1-p}\right)^{k}} = \left(\dfrac{1-q}{1-p}\right)$$

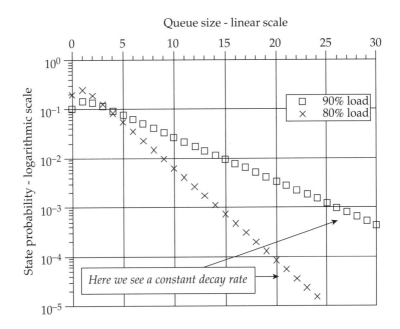

Figure 14.2. The Decay Rate of the State Probabilities for the M/D/1 Queueing System

But, as we mentioned previously, this is not true for most queueing systems. A good example of how the decay rate takes a little while to settle down can be found in the state probabilities generated using the analysis, developed in Chapter 7, for an output buffer. Let's take the case in which the number of arriving cells per time slot is Poisson-distributed, i.e. the M/D/1, and choose an arrival rate of 0.9 cells per time slot. The results are shown in Table 14.1.

The focus of buffer analysis in packet-based networks is always to evaluate probabilities associated with information loss and delay. For this reason we concentrate on the state probabilities *as seen by an arriving*

Table 14.1. Change in Decay Rate for M/D/1 with 90% Load

Radio	DR
$s(1)/s(0)$	1.4596
$s(2)/s(1)$	0.9430
$s(3)/s(2)$	0.8359
$s(4)/s(3)$	0.8153
$s(5)/s(4)$	0.8129
$s(6)/s(5)$	0.8129
$s(7)/s(6)$	0.8129

packet. This is in contrast to those *as seen by a departing packet*, as in classical queueing theory, or *as left at random instants* as we used in the time-slotted ATM buffer analysis of Chapter 7. The key idea is that, by finding the probability of what is seen ahead of an arriving packet, we have a very good indicator of both:

- the waiting time – i.e. the sum of the service time of all the packets ahead in the queue
- the loss – the probability that the buffer overflows a finite length is often closely approximated by the probability that the infinite buffer model contains more than would fit in the given finite buffer length.

Using the decay rate to approximate the buffer overflow probability

Having a constant decay rate is just the same as saying that we have a geometric progression for the state probabilities:

$$\Pr\{k\} = (1 - p) \cdot p^k$$

To find the tail probability, i.e. the probability associated with values greater than k, we have

$$\Pr\{>k\} = 1 - \Pr\{0\} - \Pr\{1\} - \cdots - \Pr\{k\}$$

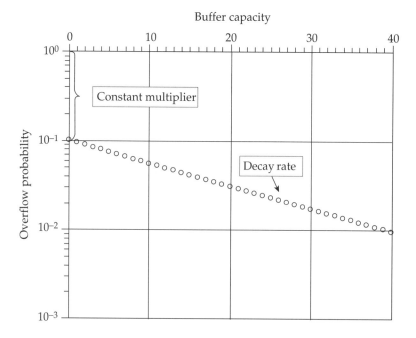

Figure 14.3.   Decay Rate Offset by a Constant Multiplier

After substituting in the geometric distribution, and doing some algebraic manipulation we have

$$\Pr\{>k\} = 1 - (1-p) - (1-p) \cdot p - \cdots - (1-p) \cdot p^k$$

$$p \cdot \Pr\{>k\} = p - (1-p) \cdot p - \cdots - (1-p) \cdot p^k - (1-p) \cdot p^{k+1}$$

$$(1-p) \cdot \Pr\{>k\} = 1 - (1-p) - p + (1-p) \cdot p^{k+1}$$

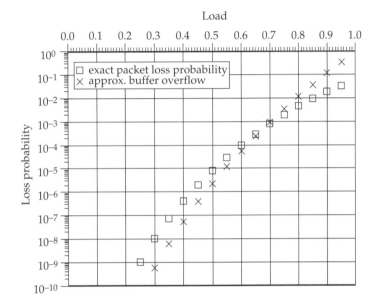

Figure 14.4. Comparison of $Q(x)$ and Loss Probability for the M/D/1 Queue Model, with a Finite Buffer Capacity of 10 Packets

which, after dividing through by $(1 - p)$, yields

$$\Pr\{>k\} = p^{k+1}$$

However, many of the buffer models we have to deal with are best modelled by a constant decay rate, d_r, that is offset by another constant multiplier, C_m. This is illustrated in Figure 14.3. If we know both the value of the decay rate and the constant multiplier then we can estimate the buffer overflow probability from:

$$\Pr\{\text{buffer overflow}\} \approx C_m \cdot d_r^{X+1}$$

where X is the buffer length in packets.

This ties in very nicely with our earlier work in Chapter 9 on the burst-scale loss and burst-scale delay models for ATM buffering, where the value of the constant was evaluated via the probability that 'a cell needs a buffer'. We use a similar approach in Chapter 15 for evaluating the resource implications of many ON–OFF sources being multiplexed in an IP router.

At this stage it is worth looking at some numerical comparisons for typical queueing systems, plotting the probability of buffer over-flow against the packet loss probability. Figure 14.4 compares these two measures for the M/D/1 system. This comparison (i.e. using state probabilities seen by arrivals) shows that $\Pr\{\text{infinite buffer contains} > X\}$ is a good approximation for the loss probability. This is the sort of simplification that is frequently exploited in buffering analyses.

BALANCE EQUATIONS FOR PACKET BUFFERING: EXCESS-RATE QUEUEING ANALYSIS

The advantage of the Geo/Geo/1 is that it is simple, and the high variability in its service times have allowed some to claim it is a 'worst-case' model. We need to note two points: the Bernoulli input process assumes arrivals as an instantaneous batch (which, as we will see in the next chapter, has a significant effect); and the geometric distribution of the packet lengths is an overestimation of the amount of variation likely to be found in real IP networks. The second of these problems, that the geometric is an unrealistic model of IP packets as it gives no real upper limit on packet lengths, can be overcome by more realistic packet-size distributions.

To address this, we develop an analytical result into which a variety of different packet-size distributions can be substituted relatively simply. To begin with, we assume fixed-size packets (i.e. the M/D/1 queue) and derive a formula that is more convenient to use than the recurrence equations of Chapter 7 and significantly more accurate than the heavy-traffic approximations of Chapter 8. This formula can be applied to cell-scale queueing in ATM as well as to packet queueing for real-time services such as voice-over-IP (which have fixed, relatively short, packet lengths).

Then we show how this formula can be applied for variable-size packets of various distributions. One particular distribution of interest is the bi-modal case: here, the packet lengths take one of two values, either the shortest possible or the longest possible. The justification for this is that in real IP networking situations there is a clear division of packets along these lines; control packets (e.g. in RSVP and TCP) tend to be very short, and data packets tend to be the maximum length allowable for the underlying sub-network technology.

The excess-rate M/D/1, for application to voice-over-IP

We introduced the notion of 'excess-rate' arrivals in Chapter 9 when we considered burst-scale queueing behaviour. Then, we were looking at the excess of arrival rate over service rate for durations of milliseconds or more, i.e. multiple cell slots. In the example of Figure 9.2, the excess rate was 8% of the service capacity over a period of 24 cell slots, i.e. approx. 2 cells in 68 μs. Typical bursts last for milliseconds, and so if this excess rate lasts for 2400 time slots, then about 200 cells must be held temporarily in the output buffer, or they are lost if there is insufficient buffer space.

Now, suppose we reduce the duration over which we define excess-rate arrivals to the time required to serve a fixed-length packet. Let this duration be our fundamental unit of time. Thus, 'excess-rate' (ER) packets are those which must be buffered as they represent an excess of *instantaneous* arrival rate over the service rate; if N packets arrive in any time unit, then that time unit experiences $N - 1$ excess packets.

Why should we do this? Well, for two reasons. First, we get a clearer idea of how the queue changes in size: for every excess-rate packet, the queue increases by one; a single packet arrival causes no change in the queue state (because one packet is also served), and the queue only decreases when there are no packets arriving in any time unit (see Figure 14.5). We can then focus on analysing the behaviour that causes the queue to change in size. Instead of connecting 'end of slot k' with 'end of slot $k + 1$' via a balance equation (i.e. using an Imbedded Markov Chain at ends of slots, as in Chapter 7), we connect the arrival of excess-rate packets via the balance equations (in a similar way to the discrete fluid-flow approach in Chapter 9).

Secondly, it gives us the opportunity to use a form of arrival process which simplifies the analysis. We alter the Poisson process to produce a geometric series for the tail of the distribution (Figure 14.6), giving a geometrically distributed number of ER packets per time unit. We call

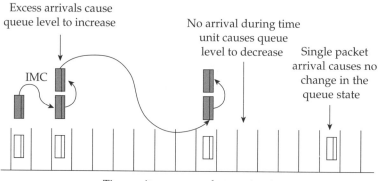

Excess arrivals cause queue level to increase

No arrival during time unit causes queue level to decrease

Single packet arrival causes no change in the queue state

IMC

Time unit = mean packet service time

Figure 14.5. Excess-Rate Queueing Behaviour

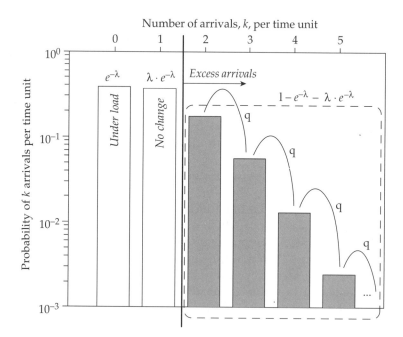

Figure 14.6. The Geometrically Approximated Poisson Process

this the Geometrically Approximated Poisson Process (GAPP) and define the conditional probability

$$q = \text{Pr}\{\text{another ER packet arrives in a time unit}|\text{just had one}\}$$

Thus the mean number of ER packets in an excess-rate batch, $E[B]$, is given by

$$E[B] = \frac{1}{1-q}$$

But we can find an expression for $E[B]$ based on the arrival probabilities:

$$E[B] = \frac{\sum_{i=2}^{\infty}(i-1)\cdot a(i)}{1-a(0)-a(1)}$$

where

$$a(k) = \Pr\{k \text{ arrivals in a packet service time}\}$$

The numerator weights all the probabilities of having i packets arriving by the number that are actually excess-rate packets, i.e. $i-1$. This ranges over all situations in which there is at least one excess-rate packet arrival. The denominator normalizes the probabilities to this condition (that there are ER arrivals). A simple rearrangement of the numerator gives

$$E[B] = \frac{\sum_{i=1}^{\infty} i \cdot a(i) - \sum_{i=1}^{\infty} a(i)}{1-a(0)-a(1)} = \frac{E[a]-(1-a(0))}{1-a(0)-a(1)}$$

where $E[a]$ is the mean number of packets arriving per unit time. We now have an expression for the parameter of the geometric ER series:

$$q = 1 - \frac{1}{E[B]} = 1 - \frac{1-a(0)-a(1)}{E[a]-1+a(0)}$$

Consider now how the queue increases in size by one packet. We define the state probability as

$$p(k) = \Pr\{\text{an arriving excess-rate packet finds } k \text{ packets in the buffer}\}$$

Remember that we are creating an Imbedded Markov Chain at excess-rate arrival instants. Thus to move from state k to $k+1$ either we need another ER packet in the same service time interval, with probability q, or for the queue content to remain unchanged until the next ER packet arrival. To express this latter probability we need to define

$$d(k) = \Pr\{\text{queue content decreases by } k \text{ packets between ER arrivals}\}$$

and

$$D(k) = \Pr\{\text{queue content decreases by at least } k \text{ packets between ER arrivals}\}$$

The queue size decreases only when there is no arrival, and it remains the same only when there is one arrival. For the queue to decrease by k packets between excess arrivals, there must be k slots with no arrivals, with probability $a(0)$, any number of slots with one arrival, with probability $a(1)$, and then a slot with excess arrivals, with probability $1 - a(0) - a(1)$. So $d(k)$ is given by

$$d(k) = \sum_{n=k}^{\infty} {}^{n}C_{k} \cdot (a(0))^{k} \cdot (a(1))^{n-k} \cdot (1 - a(0) - a(1))$$

In fact, we only need $d(0)$ in the queue analysis, and this is simply

$$d(0) = \sum_{n=0}^{\infty} (a(1))^{n} \cdot (1 - a(0) - a(1))$$

which reduces to

$$d(0) = \frac{(1 - a(0) - a(1))}{1 - a(1)}$$

We also require

$$D(1) = 1 - d(0) = \frac{a(0)}{1 - a(1)}$$

Now, for the balance equations: in a similar way to the discrete fluid-flow analysis in Chapter 9, we develop these by equating the up and down probabilities of crossing between adjacent queue states. As before, we are concerned with the state as seen by an excess-rate arrival, so we must consider arriving packets one at a time. Thus the state can only ever increase by one.

Initially, let the buffer capacity be X packets. To cross between states $X - 1$ and X, an arriving excess-rate packet sees $X - 1$, taking the queue state up to X, and another excess-rate packet follows to see the queue in state X. This happens either immediately, with probability q, or after any number of time units in which the queue state stays the same, i.e. with probability $(1 - q) \cdot d(0)$. So the probability of going up is

$$\Pr\{\text{going up}\} = (q + (1 - q) \cdot d(0)) \cdot p(X - 1)$$

To go down, an arriving excess-rate packet sees X in the queue and is lost (because the buffer is full) and then there is a gap containing any number of time units, at least one of which is empty and the rest in which the queue state does not change. Then the next excess-rate arrival sees fewer than X in the queue. For this to happen, it is simply the probability that the next ER packet *does not* see X, or, put another way, one minus the probability that the next ER packet *does* see X. This latter probability has the same conditions as the up transition, i.e. either another ER packet follows immediately, or it follows after any number of time units in which the queue state stays the same. So

$$\Pr\{\text{going down}\} = (1 - (q + (1 - q) \cdot d(0))) \cdot p(X)$$

Equating the probabilities of going up and down gives

$$(q + (1 - q) \cdot d(0)) \cdot p(X - 1) = (1 - (q + (1 - q) \cdot d(0))) \cdot p(X)$$

For a line between $X - 1$ and $X - 2$, equating probabilities gives

$$(q + (1 - q) \cdot d(0)) \cdot p(X - 2) = (1 - (q + (1 - q) \cdot d(0))) \cdot D(1) \cdot p(X)$$
$$+ (1 - (q + (1 - q) \cdot d(0))) \cdot D(1) \cdot p(X - 1)$$

The left-hand side is the probability of going up, and has the same conditions as before. The probability of coming down, on the right-hand side of the equation, contains two possibilities. The first term is for an arriving ER packet which sees X in the queue and is lost (because the buffer is full) and the second term is for an arriving ER packet which sees $X - 1$ in the queue, taking the state of the queue up to X. Then, in both cases, there is a period without ER packets during which the queue content decreases by at least two empty time units, so that the next ER arrival sees fewer than $X - 1$ in the queue.

Substituting for $p(X)$, and rearranging gives

$$(q + (1 - q) \cdot d(0)) \cdot p(X - 2) = D(1) \cdot p(X - 1)$$

In the general case, for a line between $X - i + 1$ and $X - i$, the probability of going up remains the same as before, i.e. the only way to go up is for an ER packet to see $X - i$, and to be followed (either immediately or after a period during which the queue state remains the same) by another ER packet which sees $X - i + 1$. The probability of going down consists of many components, one for each state above $X - i$, but they can be arranged in two groups: the probability of coming down from $X - i + 1$ itself; and the probability of coming down to below $X - i + 1$ from above $X - i + 1$. This latter is just the probability of going down between $X - i + 2$ and $X - i + 1$ multiplied by $D(1)$, which is the same as going up from $X - i + 1$ multiplied by $D(1)$. This is precisely the same grouping as illustrated in Figure 9.10 for the discrete fluid-flow approach.

The general equation then is

$$(q + (1 - q) \cdot d(0)) \cdot p(X - i) = D(1) \cdot p(X - i + 1)$$

so

$$p(X) = p(0) \cdot \left[\frac{(q + (1-q) \cdot d(0))}{D(1)} \right]^X$$

The state probabilities must sum to 1, so

$$\sum_{i=0}^{X} p(i) = p(0) \cdot \sum_{i=0}^{X} \left[\frac{(q + (1-q) \cdot d(0))}{D(1)} \right]^i = 1$$

and we can find $p(0)$ thus

$$p(0) = \frac{1 - \dfrac{(q + (1-q) \cdot d(0))}{D(1)}}{1 - \left[\dfrac{(q + (1-q) \cdot d(0))}{D(1)} \right]^{X+1}}$$

Now, although we have assumed a finite buffer capacity of X packets, let us now assume $X \to \infty$. The term in the denominator for $p(0)$ tends to 1, and so the state probabilities can be written

$$p(k) = \left(1 - \frac{(q + (1-q) \cdot d(0))}{D(1)}\right) \cdot \left[\frac{(q + (1-q) \cdot d(0))}{D(1)} \right]^k$$

Substituting for q, $d(0)$ and $D(1)$ gives

$$p(k) = \left(1 - \frac{E[a] \cdot (1 - a(1)) - 1 + a(1) + (a(0))^2}{a(0) \cdot (E[a] - 1 + a(0))}\right)$$

$$\cdot \left[\frac{E[a] \cdot (1 - a(1)) - 1 + a(1) + (a(0))^2}{a(0) \cdot (E[a] - 1 + a(0))} \right]^k$$

Now, although this expression looks rather messy, its structure is quite simple:

$$p(k) = (1 - \text{decay rate}) \cdot [\text{decay rate}]^k$$

The probability that the queue exceeds k packets is then a geometric progression which, after rearrangement, yields

$$Q(k) = [\text{decay rate}]^{k+1}$$

i.e.

$$Q(k) = \left[\frac{E[a] \cdot (1 - a(1)) - 1 + a(1) + (a(0))^2}{a(0) \cdot (E[a] - 1 + a(0))} \right]^{k+1}$$

These then are the general forms for $p(k)$ and $Q(k)$, into which we can substitute appropriate expressions from the Poisson distribution, i.e.

$$E[a] = \lambda$$

$$a(0) = e^{-\lambda}$$

$$a(1) = \lambda \cdot e^{-\lambda}$$

to give

$$p(k) = \left(1 - \frac{\lambda \cdot e^{\lambda} - e^{\lambda} - \lambda^2 + \lambda + e^{-\lambda}}{\lambda - 1 + e^{-\lambda}}\right) \cdot \left[\frac{\lambda \cdot e^{\lambda} - e^{\lambda} - \lambda^2 + \lambda + e^{-\lambda}}{\lambda - 1 + e^{-\lambda}}\right]^k$$

$$Q(k) = \left[\frac{\lambda \cdot e^{\lambda} - e^{\lambda} - \lambda^2 + \lambda + e^{-\lambda}}{\lambda - 1 + e^{-\lambda}}\right]^{k+1}$$

Well, was it really worth all the effort? Let's take a look at some results. Figure 14.7 shows the queue state probabilities for three different values of $\lambda = 0.55, 0.75, 0.95$. In the figure, the lines are the exact results found using the approach developed in Chapter 7, with Poisson input, and the markers show the results from the excess-rate analysis with GAPP input. Note that the results from the exact analysis are discrete, not continuous, but are shown as continuous lines for clarity.

Figure 14.8 shows the results for $Q(k)$, the probability that the queue exceeds k, comparing exact and excess-rate GAPP analyses. Figure 14.9 compares the excess-rate GAPP results with those from the heavy-traffic analysis in Chapter 8. It is clear that the excess-rate GAPP provides a very accurate approximation to the exact results across the full range of load values, and it is significantly more accurate than the heavy-traffic approximation.

The excess-rate solution for best-effort traffic

But how can we use the excess-rate analysis if we have variable-length packets, as is typical with current best-effort IP networks? The key here is to go back to the definition of $a(k)$, the probability of k arrivals *in a packet service time*, because it is from $a(0)$, $a(1)$ and the mean of this distribution, $E[a]$, that we derive the excess-rate queueing behaviour (see Figures 14.5 and 14.6).

Previously, we assumed a constant service time, $T = 1$ time unit. Thus, for packets arriving at an average rate of λ packets per time unit, the probability that there are no packets arriving in one packet service time is given by

$$a(0) = \frac{(\lambda \cdot T)^0}{0!} \cdot e^{-\lambda \cdot T} = e^{-\lambda}$$

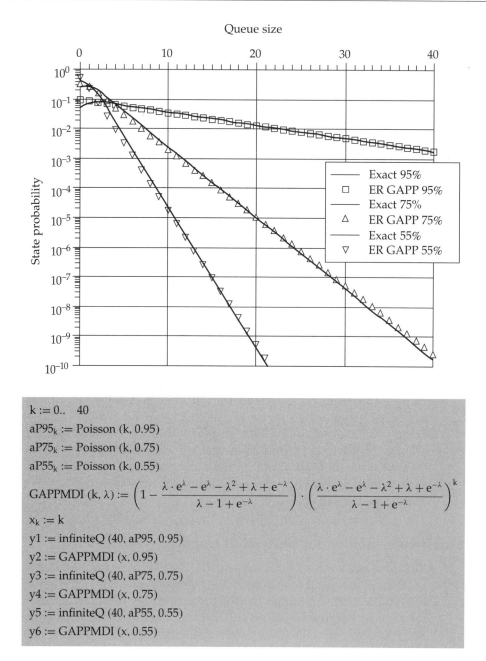

Figure 14.7.   State Probability Distributions at Various Load Levels, Comparing the Exact Analysis and Excess-Rate Analysis Methods

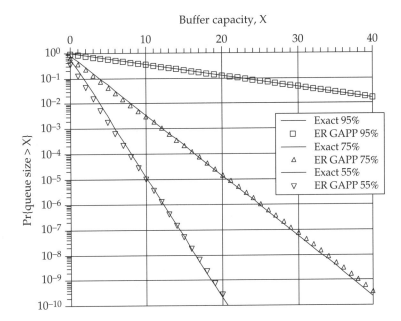

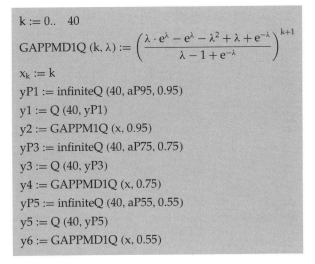

Figure 14.8. Probability that the Queue Exceeds X for Various Load Levels, Comparing the Exact Analysis and Excess-Rate Analysis Methods

But what if 50% of the packets are short, say 40 octets, and 50% of the packets are long, say 960 octets? The average packet length is 500 octets, equivalent to one time unit. The probability that there are no packets arriving in one packet service time is now a weighted sum of the two possible situations, i.e.

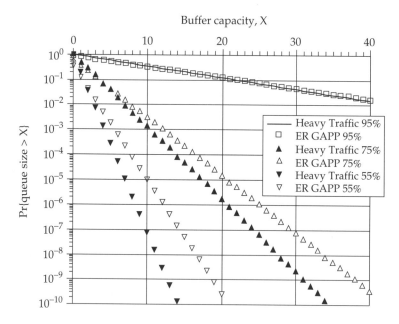

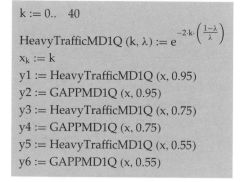

Figure 14.9. Probability that the Queue Exceeds X for Various Load Levels, Comparing the Excess-Rate and Heavy Traffic Approximation Methods

$$a(0) = 0.5 \cdot \frac{\left(\lambda \cdot \frac{40}{500}\right)^0}{0!} \cdot e^{-\lambda \cdot \frac{40}{500}} + 0.5 \cdot \frac{\left(\lambda \cdot \frac{960}{500}\right)^0}{0!} \cdot e^{-\lambda \cdot \frac{960}{500}}$$

giving

$$a(0) = 0.5 \cdot e^{-\lambda \cdot \frac{2}{25}} + 0.5 \cdot e^{-\lambda \cdot \frac{48}{25}}$$

The same approach applies for $a(1)$, i.e.

$$a(1) = 0.5 \cdot \lambda \cdot \frac{2}{25} \cdot e^{-\lambda \cdot \frac{2}{25}} + 0.5 \cdot \lambda \cdot \frac{48}{25} \cdot e^{-\lambda \cdot \frac{48}{25}}$$

Let's make this more general. Assume we have short packets of length 1 time unit, and long packets of length n time units. The proportion of short packets is p_s, and so the proportion of long packets is $(1 - p_s)$. Packets arrive at a rate of λ packets per time unit. The mean service time (in time units) is then simply

$$s = p_s \cdot 1 + (1 - p_s) \cdot n$$

and mean number of arrivals per packet service time (i.e. the utilization) is given by

$$E[a] = \rho = \lambda \cdot s = \lambda \cdot \{p_s + (1 - p_s) \cdot n\}$$

The general form for the probability of k arrivals in a packet service time is then given by

$$a(k) = p_s \cdot \frac{\lambda^k}{k!} \cdot e^{-\lambda} + (1 - p_s) \cdot \frac{(n \cdot \lambda)^k}{k!} \cdot e^{-n \cdot \lambda}$$

So for bi-modal packet-size distributions, we have the following expressions

$$E[a] = \lambda \cdot (p_s + (1 - p_s) \cdot n)$$

$$a(0) = p_s \cdot e^{-\lambda} + (1 - p_s) \cdot e^{-n \cdot \lambda}$$

$$a(1) = p_s \cdot \lambda \cdot e^{-\lambda} + (1 - p_s) \cdot n \cdot \lambda \cdot e^{-n \cdot \lambda}$$

which we can substitute into the general forms for $p(k)$ and $Q(k)$ from the previous section. Note that n does not have to be integer-valued.

It is not necessary to limit the packet sizes to two distinct values, though. The above process can be generalized further to the case of the general service-time distribution, hence providing an excess-rate solution for the M/G/1 queueing system.

$$E[a] = \lambda \cdot \sum_{i=1}^{\infty} g(i) = \lambda$$

$$a(0) = \sum_{i=1}^{\infty} g(i) \cdot e^{-i \cdot \lambda}$$

$$a(1) = \sum_{i=1}^{\infty} g(i) \cdot i \cdot \lambda \cdot e^{-i \cdot \lambda}$$

where

$$g(k) = \Pr\{a \text{ packet requires } k \text{ time units to be served}\}$$

But for now, let's keep to the bi-modal case and look at some results (in Figure 14.10) for different lengths and proportions of short and long

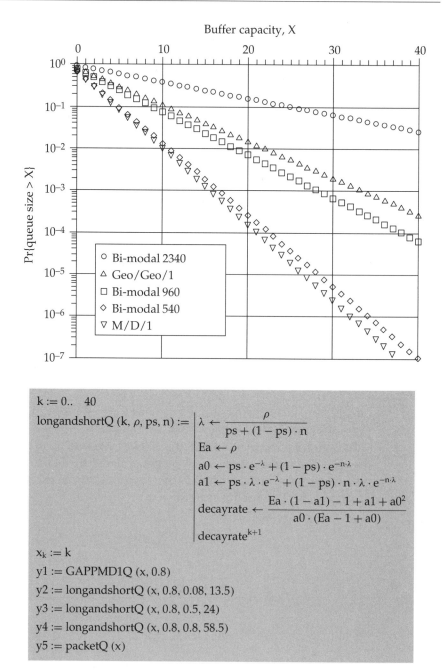

Figure 14.10. Probability that the Queue Exceeds X for Different Service Distributions (Geometric, Deterministic and Bi-modal)

packets. We fix the short packet length to 40 octets, the mean packet length to 500 octets and the load to 0.8 packets arriving per mean service time. We set the time unit to be the time to serve a short packet, so

$$s = \frac{500}{40} = p_s \cdot 1 + (1 - p_s) \cdot n$$

Figure 14.10 shows results for three different lengths for the long packets: 2340, 960, and 540 octets. These give values of $n = 58.5$, 24 and 13.5 with corresponding values of $p_s = 0.8$, 0.5 and 0.08 respectively. Also in the figure are the results for the M/D/1 with the same load of 0.8, and the Geo/Geo/1 results from Figure 14.1 (a load of 0.8 and mean packet-size of 500 octets). Note that the M/D/1 gives a lower bound on the probabilities, but the Geo/Geo/1 is *not the worst case*. Introducing a small proportion of short packets, and hence slightly increasing the length of the long packets (to maintain an average of 500 octets) results in a decay rate only a little higher than the M/D/1. When there are equal proportions of short and long packets, the decay rate approaches that for the Geo/Geo/1. However, when most of the packets are short, and only 20% are (very) long packets, the decay rate is rather worse than that for the Geo/Geo/1 queue.

15 Resource Reservation

go with the flow

QUALITY OF SERVICE AND TRAFFIC AGGREGATION

In recent years there have been many different proposals (such as Integrated Services [15.1], Differentiated Services [15.2], and RSVP [15.3]) for adding quality of service (QoS) support to the current best-effort mode of operation in IP networks. In order to provide guaranteed QoS, a network must be able to anticipate traffic demands, assess its ability to supply the necessary resources, and act either to accept or reject these demands for service. This means that users must state their communications requirements in advance, in some sort of service request mechanism. The details of the various proposals are outside the scope of this book, but in this chapter we analyse the key queueing behaviours and performance characteristics underlying the resource assessment.

To be able to predict the impact of new demands on resources, the network needs to record state information. Connection-orientated technologies such as ATM record per-connection information in the network as 'hard' state. This information must be explicitly created for the duration of the connection, and removed when no longer needed. An alternative approach (adopted in RSVP) is 'soft' state, where per-flow information is valid for a pre-defined time interval, after which it needs to be 'refreshed' or, if not, it lapses.

Both approaches, though, face the challenge of scalability. Per-flow or per-connection behaviour relates to individual customer needs. With millions of customers, each one initiating many connections or flows, it is important that the network can handle these efficiently, whilst still providing guaranteed QoS. This is where traffic aggregation comes in. ATM technology introduces the concept of the virtual path – a bundle of virtual channels whose cells are forwarded on the basis of their VPI value only. In IP, packets are classified into behaviour aggregates, identified by a field in the IP header, and forwarded and queued on the basis of the value of that field.

In this chapter, we concentrate on characterizing these traffic aggregates, and analysing their impact on the network to give acceptable QoS for the end users. Indeed, our approach divides into these two stages: aggregation, and analysis (using the excess-rate analysis from Chapter 9).

CHARACTERIZING AN AGGREGATE OF PACKET FLOWS

In the previous chapter, we assumed that the arrival process of packets could be described by a Poisson distribution (which we modified slightly, to derive accurate results for both M/D/1 and M/G/1 queueing systems). This assumption allowed for multiple packets, from different input ports, to arrive simultaneously (i.e. within one packet service time) at an output port, and hence require buffering. This is a valid assumption when the input and output ports are of the same speed (bit-rate) and there is no correlation between successive arrivals on an input port.

However, if the input ports are substantially slower than the output port (e.g. in a typical access multiplexing scenario), or packets arrive in bursts at a rate slower than that allowed by the input port rate (within the core network), then the Poisson assumption is less valid. Why? Well, suppose that the output port rate is 1000 packet/s and the traffic on the input port is limited to 100 packet/s (either because of a physical bit-rate limit, or because of the packet scheduling at the previous router). The minimum time between arrivals from any single input port is then 10 ms, during which time the output port could serve up to 10 packets. The Poisson assumption allows for arrivals during any of the 10 packet service times, but the actual input process does not.

So, we characterize these packet flows as having a mean duration, T_{on}, and an arrival rate when active, h (packet/s). Thus each flow comprises $T_{on} \cdot h$ packets, on average. If the overall mean load is A_p packet/s, then the rate of flows arriving is simply

$$F = \frac{A_p}{T_{on} \cdot h}$$

We can interpret this arrival process in terms of erlangs of offered traffic:

$$\text{offered traffic} = \frac{A_p}{h} = F \cdot T_{on}$$

i.e. the flow attempt rate multiplied by the mean flow duration.

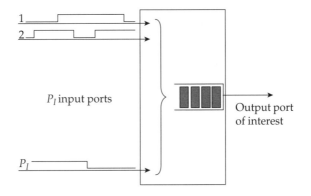

Figure 15.1. Access Multiplexor or Core Router

It may be that there is a limit on the number of input ports, P_I, sending flows to the particular output port of interest (see Figure 15.1). In this case, the two scenarios (access multiplexor, or core router/switch) differ in terms of the maximum number of flows, N, at the output port. For the access multiplexor, with slow speed input ports of rate h packet/s, the maximum number of simultaneous flows is

$$N = P_I$$

However, for the core router with input port speeds of C packet/s, the maximum possible number of simultaneous flows it can support is

$$N = P_I \cdot \left\lfloor \frac{C}{h} \right\rfloor$$

i.e. each input port can carry multiple flows, each of rate h, which have been multiplexed together upstream of this router.

PERFORMANCE ANALYSIS OF AGGREGATE PACKET FLOWS

The first task is to simplify the traffic model, comprising N input sources, to one in which there is a single aggregate input process to the buffer (see Figure 15.2), thus reducing the state space from 2^N possible states to just 2. This aggregate process is either in an ON state, in which the input rate exceeds the output rate, or in an OFF state, when the input rate is not zero, but is less than the output rate.

For the aggregate process, the mean rate in the ON state is denoted R_{on}, and in the OFF state is R_{off}. When the aggregate process is in the ON state, the total input rate exceeds the service rate, C, of the output port, and the buffer fills:

$$\text{rate of increase} = R_{on} - C$$

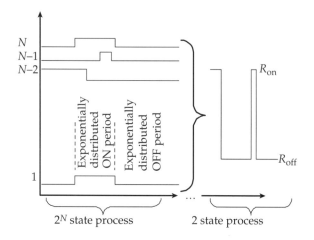

Figure 15.2.   State Space Reduction for Aggregate Traffic Process

The average duration of this period of increase is denoted $T(on)$. To be in the ON state, more than C/h sources must be active. Otherwise the aggregate process is in the OFF state. This is illustrated in Figure 15.3.

In the OFF state, the total input rate is less than the service rate of the output port, so, allowing the buffer to empty,

$$\text{rate of decrease} = C - R_{off}$$

The average duration of this period of decrease is denoted $T(off)$.

Reducing the system in this manner has obvious attractions; however, just having a simplifying proposal does not lead directly to the model in detail. Specifically, we need to find values for the four parameters in our two-state model, a process which is called 'parameterization'.

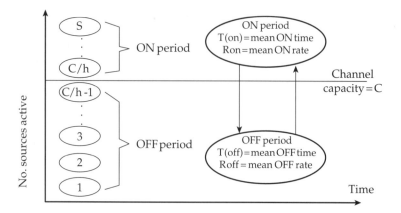

Figure 15.3.   Two-State Model of Aggregate Packet Flows

Parameterizing the two-state aggregate process

Consider the left-hand side of Figure 15.3. Here we show the combined input rates, depending on how many packet flows are active. The capacity assigned to this traffic aggregate is C packet/s – this may be the total capacity of the output port, or just a fraction if there is, for example, a weighted fair queue scheduling scheme in operation. If C/h packet flows are active, then the input and output rates of the queue are equal, and the queue size remains constant. From the burst-scale point of view, the queue is constant, although there will be small-scale fluctuations due to the precise timing of packet arrival and departure instants. If more packet flows are active, the queue increases in size because of the excess rate; with fewer packet flows active, the queue decreases in size.

Let us now view the queueing system from the point of view of the arrival and departure of *packet flows*. The maximum number of packet flows that can be served simultaneously is

$$N_0 = \frac{C}{h}$$

We can therefore think of the output port as having N_0 servers and a buffer for packet flows which are waiting to be served. If we can find the mean number waiting to be served, given that there are some waiting, we can then calculate the mean rate in the ON state, R_{on}, as well as the mean duration in the ON state, $T(on)$.

Assuming a memoryless process for the arrival of packet flows (a reasonable assumption, since flows are typically triggered by user activity), this situation is then equivalent to the system modelled by Erlang's waiting-call analysis. Packet flows are equivalent to calls, the output port is equivalent to N_0 circuits, and we assume infinite waiting space. The offered traffic, in terms of packet flows, is given by

$$A = \frac{A_p}{h} = F \cdot T_{on}$$

Erlang's waiting-call formula gives the probability of a call (packet flow) being delayed as

$$D = \frac{\left\{ \dfrac{A^{N_0}}{N_0!} \cdot \left(\dfrac{N_0}{N_0 - A} \right) \right\}}{\left\{ \displaystyle\sum_{r=0}^{N_0-1} \dfrac{A^r}{r!} + \dfrac{A^{N_0}}{N_0!} \cdot \left(\dfrac{N_0}{N_0 - A} \right) \right\}}$$

or, alternatively, in terms of Erlang's loss probability, B, we have

$$D = \frac{N_0 \cdot B}{N_0 - A + A \cdot B}$$

The mean number of calls (packet flows) waiting, averaged over all calls, is given by

$$w = D \cdot \frac{A}{N_0 - A}$$

But what we need is the mean number waiting, conditioned on there being some waiting. This is simply given by

$$\frac{w}{D} = \frac{A}{N_0 - A}$$

Thus, when the aggregate traffic is in the ON state, i.e. there are some packet flows 'waiting', then the mean input rate to the output port exceeds the service rate. This excess rate is simply the product of the conditional mean number waiting and the packet rate of a packet flow, h. So

$$R_{on} = C + h \cdot \frac{A}{N_0 - A} = C + h \cdot \frac{A_p}{C - A_p}$$

The mean duration in the excess-rate (ON) state is the same as the conditional mean delay for calls in the waiting-call system. From Little's formula, we have

$$w = F \cdot t_w = \frac{A}{T_{on}} \cdot t_w$$

which, on rearranging and substituting for w, gives

$$t_w = \frac{T_{on}}{A} \cdot w = \frac{T_{on}}{A} \cdot D \cdot \frac{A}{N_0 - A}$$

So, the conditional mean delay is

$$T(on) = \frac{t_w}{D} = \frac{T_{on}}{N_0 - A} = \frac{h \cdot T_{on}}{C - A_p}$$

This completes the parameterization of the ON state. In order to parameterize the OFF state we need to make use of D, the probability that a packet flow is delayed. This probability is, in fact, the probability that the aggregate process is in the ON state, which is the long-run proportion of time in the ON state. So we can write

$$\frac{T(on)}{T(on) + T(off)} = D$$

which, after rearranging, gives

$$T(off) = T(on) \cdot \frac{1 - D}{D}$$

The mean load, in packet/s, is the weighted sum of the rates in the ON and OFF states, i.e.

$$A_p = D \cdot R_{on} + (1 - D) \cdot R_{off}$$

and so

$$R_{off} = \frac{A_p - D \cdot R_{on}}{1 - D}$$

Analysing the queueing behaviour

We have now aggregated the Poisson arrival process of packet flows into a two-state ON–OFF process. This is very similar to the ON–OFF source model in the discrete fluid-flow approach presented in Chapter 9, except that the OFF state now has a non-zero arrival rate associated with it. In the ON state, we assume that there are a geometrically distributed number of excess-rate packet arrivals. In the OFF state, we assume that there are a geometrically distributed number of free periods in which to serve excess-rate packets. Thus the geometric parameters a and s are given by

$$a = 1 - \frac{1}{T(on) \cdot (R_{on} - C)}$$

and

$$s = 1 - \frac{1}{T(off) \cdot (C - R_{off})}$$

For a finite buffer size of X, we had the following results from Chapter 9:

$$p(X - 1) = \frac{1 - a}{a} \cdot p(X)$$

and

$$p(X - i) = \frac{s}{a} \cdot p(X - i + 1)$$

The state probabilities, $p(k)$, form a geometric progression which can be written as

$$p(k) = \begin{cases} \left(\dfrac{a}{s}\right)^k \cdot p(0) & 0 < k < X \\[3mm] \left(\dfrac{s}{1-a}\right) \cdot \left(\dfrac{a}{s}\right)^k \cdot p(0) & k = X \end{cases}$$

These state probabilities must sum to 1, and so, after some rearrangement, we can find $p(0)$ thus:

$$p(0) = \frac{1 - \dfrac{a}{s}}{1 - \left(\dfrac{1-s}{1-a}\right) \cdot \left(\dfrac{a}{s}\right)^X}$$

Now, although we have assumed a finite buffer capacity of X packets for this excess-rate analysis, let us now assume $X \to \infty$. The term in the denominator for $p(0)$ tends to 1, and so the state probabilities can be written

$$p(k) = \left(1 - \frac{a}{s}\right) \cdot \left(\frac{a}{s}\right)^k$$

As we found in the previous chapter for this form of expression, the probability that the queue exceeds k packets is then a geometric progression, i.e.

$$Q(k) = \left(\frac{a}{s}\right)^{k+1}$$

This result is equivalent to the burst-scale delay factor – it is the probability that excess-rate packets see more than k in the queue. It is in our, now familiar, decay rate form, and provides an excellent approximation to the probability that a finite buffer of length k overflows. This latter is a good approximation to the loss probability.

However, we have not quite finished. We now need an expression for the probability that a packet *is* an excess-rate arrival. In the discrete fluid-flow model of Chapter 9, this was simply $(R - C)/R$ – the proportion of arrivals that are excess-rate arrivals. This simple expression needs to be modified because when the aggregate process is in the OFF state, packets are still arriving at the queue.

We need to find the ratio of the mean excess rate to the mean arrival rate. If we consider a single ON–OFF cycle of the aggregate model, then this ratio is the mean number of excess packets in an ON period to the mean number of packets arriving in the ON–OFF cycle. Thus

$$\Pr\{\text{packet is excess-rate arrival}\} = \frac{(R_{on} - C) \cdot T(on)}{A_p \cdot (T(on) + T(off))}$$

which, after substituting for R_{on}, $T(on)$ and $T(off)$, gives

$$\Pr\{\text{packet is excess-rate arrival}\} = \frac{h \cdot D}{C - A_p}$$

The queue overflow probability is then given by the expression

$$Q(x) = \frac{h \cdot D}{C - A_p} \cdot \left(\frac{1 - \dfrac{1}{T(on) \cdot (R_{on} - C)}}{1 - \dfrac{1}{T(off) \cdot (C - R_{off})}} \right)^{x+1}$$

VOICE-OVER-IP, REVISITED

In the last chapter we looked at the excess-rate M/D/1 analysis as a suitable model for voice-over-IP. The assumption of a deterministic server is reasonable, given that voice packets tend to be of fixed size, and the Poisson arrival process is a good limit for N CBR sources when N is large (as we found in Chapter 8). But if the voice sources are using activity detection, then they do not send packets during silent periods. Thus we have ON–OFF behaviour, which can be viewed as a series of overlapping packet flows (see Figure 15.1).

Suppose we have $N = 100$ packet voice sources, each producing packets at a rate of $h = 167$ packet/s, when active, into a buffer of size $X = 100$ packets and service capacity $C = 7302.5$ packet/s. The mean time when active is $T_{on} = 0.35$ seconds and when inactive is $T_{off} = 0.65$ second, thus each source has, on average, one active period every $T_{on} + T_{off} = 1$ second. The rate at which these active periods arrive, from the population of N packet sources, is then

$$F = \frac{N}{T_{on} + T_{off}} = 100 \text{ s}^{-1}$$

Therefore, we can find the overall mean load, A_p, and the offered traffic, A, in erlangs.

$$A_p = F \cdot T_{on} \cdot h = 100 \times 0.35 \times 167 = 5845 \text{ packet/s}$$

$$A = F \cdot T_{on} = 100 \times 0.35 = 35 \text{ erlangs}$$

and the maximum number of sources that can be served simultaneously, without exceeding the buffer's service rate is

$$N_0 = \frac{C}{h} = 43.728$$

which needs to be rounded down to the nearest integer, i.e. $N_0 = 43$. Let's now parameterize the two-state excess-rate model.

$$B = \frac{\dfrac{A^{N_0}}{N_0!}}{\displaystyle\sum_{r=0}^{N_0} \frac{A^r}{r!}} = 0.028\,14$$

$$D = \frac{N_0 \cdot B}{N_0 - A + A \cdot B} = 0.134\,66$$

$$R_{on} = C + h \cdot \frac{A_p}{C - A_p} = 7972.22$$

$$R_{off} = \frac{A_p - D \cdot R_{on}}{1 - D} = 5513.98$$

$$T(on) = \frac{h \cdot T_{on}}{C - A_p} = 0.0401$$

$$T(off) = T(on) \cdot \frac{1 - D}{D} = 0.257\,71$$

We can now calculate the geometric parameters, a and s, and hence the decay rate.

$$a = 1 - \frac{1}{T(on) \cdot (R_{on} - C)} = 0.962\,77$$

$$s = 1 - \frac{1}{T(off) \cdot (C - R_{off})} = 0.997\,83$$

$$\text{decay rate} = \frac{a}{s} = 0.964\,86$$

The probability that a packet is an excess-rate arrival is then

$$\Pr\{\text{packet is excess-rate arrival}\} = \frac{h \cdot D}{C - A_p} = 0.015\,43$$

and the packet loss probability is estimated by

$$Q(X) = \frac{h \cdot D}{C - A_p} \cdot \left(\frac{a}{s}\right)^{X+1} = 4.161\,35 \times 10^{-4}$$

Figure 15.4 shows these analytical results on a graph of $Q(x)$ against x. The Mathcad code to generate the analytical results is shown in Figure 15.5. Also shown, as a dashed line in Figure 15.4, are the results of applying the burst-scale analysis (both loss and delay factors, from Chapter 9) to the same scenario. Simulation results for this scenario show a decay rate of approximately 0.97. The figure of 0.964 86 obtained from the excess-rate aggregate flow analysis is very close to these simulation results, and illustrates the accuracy of the excess-rate technique. In contrast, the burst-scale delay factor gives a decay rate of 0.998 59. This latter is typical of other published techniques which tend to overestimate the decay rate by a significant margin; the interested reader is referred to [15.4] for a more comprehensive comparison.

If we return to the M/D/1 scenario, where we assume that the voice sources are of a constant rate, how many sources can be supported over the same buffer, and with the same packet loss probability? The excess-rate analysis gives us the following equation:

$$Q(100) = \left[\frac{\lambda \cdot e^{\lambda} - e^{\lambda} - \lambda^2 + \lambda + e^{-\lambda}}{\lambda - 1 + e^{-\lambda}}\right]^{101} = 4.161\,35 \times 10^{-4}$$

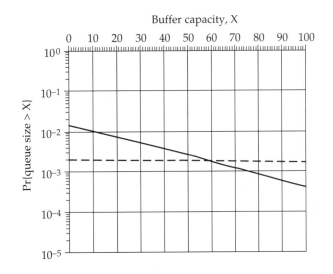

Figure 15.4. Packet Loss Probability Estimate for Voice Sources, Based on Excess-Rate Aggregate Flow Analysis

$k := 0..\ 100$

$\mathrm{afQ}\,(k\,,h\,,\mathrm{Tflow}\,,\mathrm{Ap},\mathrm{C}\,) := \bigg|\, A \leftarrow \dfrac{\mathrm{Ap}}{h}$

$\qquad N0 \leftarrow \mathrm{floor}\left(\dfrac{C}{h}\right)$

$\qquad B \leftarrow \dfrac{A^{N0}}{N0!} \cdot \dfrac{1}{\sum_{r=0}^{N0} \dfrac{A^{r}}{r!}}$

$\qquad D \leftarrow \dfrac{N0 \cdot B}{N0 - A + A \cdot B}$

$\qquad \mathrm{Ton} \leftarrow \dfrac{h \cdot \mathrm{Tflow}}{C - \mathrm{Ap}}$

$\qquad \mathrm{Toff} \leftarrow \mathrm{Ton} \cdot \left(\dfrac{1 - D}{D}\right)$

$\qquad \mathrm{Ron} \leftarrow C + h \cdot \dfrac{\mathrm{Ap}}{C - \mathrm{Ap}}$

$\qquad \mathrm{Roff} \leftarrow \dfrac{\mathrm{Ap} - D \cdot \mathrm{Ron}}{1 - D}$

$\qquad \mathrm{decayrate} \leftarrow \dfrac{1 - \dfrac{1}{\mathrm{Ton} \cdot (\mathrm{Ron} - C)}}{1 - \dfrac{1}{\mathrm{Toff} \cdot (C - \mathrm{Roff})}}$

$\qquad \mathrm{probexcess} \leftarrow \dfrac{h \cdot D}{C - \mathrm{Ap}}$

$\qquad \mathrm{probexcess} \cdot \mathrm{decayrate}^{k+1}$

$x_k := k$

$y := \mathrm{afQ}\,(x\,, 167\,, 0.35\,, 5845\,, 7302.5\,)$

Figure 15.5. Mathcad Code for Excess-Rate Aggregate Flow Analysis

which is plotted in Figure 15.6 for values of load ranging from 0.8 up to 0.99 of the service capacity.

The value of loss probability we require occurs at an offered load of about 0.96; in fact 0.961 yields a loss probability of 4.164×10^{-4}. This offered load is just the total input rate from all the CBR sources, divided by the service rate of the output port. So, we have

$$\frac{N_{CBR} \cdot h}{C} = 0.961$$

$$N_{CBR} = \frac{0.961 \times 7302.5}{167} = 42.02$$

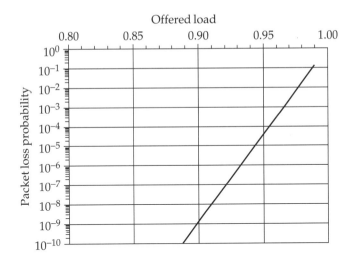

Figure 15.6. Packet Loss Probability Estimate for Voice Sources, Based on Excess-Rate M/D/1 Analysis

Thus, instead of transporting 100 voice sources, we can only carry 42 if there is no activity detection. This gives us nearly two and half times increase in loading efficiency when activity detection is used.

TRAFFIC CONDITIONING OF AGGREGATE FLOWS

Within both the Integrated Services and Differentiated Services architectures [15.1, 15.2], the concept of a token bucket is introduced to describe the load imposed by either individual, or aggregate, flows. The token bucket is, in essence, the same as the leaky bucket used in ATM usage parameter control, which we described in Chapter 11. It is normally viewed as a *pool*, of capacity B octet tokens, being filled at a rate of R octet token/s. If the pool contains enough tokens for the arriving packet, then the packet is sent on into the network, and the token bucket is drained of the appropriate number of octet tokens. However, if there are insufficient tokens, then the packet is either discarded, or marked as best-effort, or delayed until enough tokens have replenished the bucket.

In both architectures, the token bucket can be used to define a traffic profile, and hence police traffic flows (either single or aggregate). As we found in Chapter 11, we can analyse the leaky (or token) bucket as a buffer, even though in reality it is not used to delay packets. So, if we have an aggregate flow, the same analysis used to assess queueing performance can be used to dimension the token bucket.

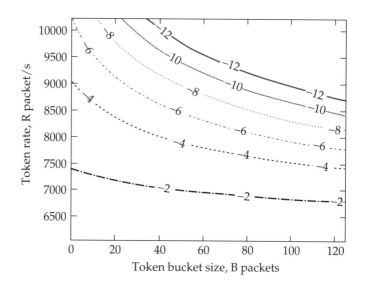

Figure 15.7. Example of Relationship between Token Bucket Parameter Values for Voice-over-IP Aggregate Traffic

Figure 15.7 shows the relationship between B and R for various values of the packet loss probability estimate (10^{-2} down to 10^{-12}). The scenario is the aggregate flow of voice-over-IP traffic, using the parameter values and formulas in the previous section. The tokens are equivalent to packets, rather than octets, in this figure. A simple scaling factor (the number of octets per packet) can be applied to convert to octets. There is a clear trade-off between rate allocation (R) and burstiness (B) for the aggregate flow. With a smaller rate allocation, the aggregate flow exceeds this value more often, and so a larger token bucket is required to accommodate these bursts.

16 IP Buffer Management

packets in the space – time continuum

FIRST-IN FIRST-OUT BUFFERING

In the chapters on packet queueing, we have so far only considered queues with first-in-first-out (FIFO) scheduling. This approach gives all packets the same treatment: packets arriving to a buffer are placed at the back of the queue, and have to wait their turn for service, i.e. after all the other packets already in the queue have been served. If there is insufficient space in the buffer to hold an arriving packet, then it is discarded.

In Chapter 13, we considered priority control in ATM buffers, in terms of space priority (access to the waiting space) and time priority (access to the server). These mechanisms enable end-to-end quality-of-service guarantees to be provided to different types of traffic in an integrated way. For IP buffer management, similar mechanisms have been proposed to provide QoS guarantees, improved end-to-end behaviour, and better use of resources.

RANDOM EARLY DETECTION – PROBABILISTIC PACKET DISCARD

One particular challenge of forwarding best-effort packet traffic is that the transport-layer protocols, especially TCP, can introduce unwelcome behaviour when the network (or part of it) is congested. When a TCP connection loses a packet in transit (e.g. because of buffer overflow), it responds by entering the slow-start phase which reduces the load on the network and hence alleviates the congestion. The unwelcome behaviour arises when *many* TCP connections do this at around the same time. If a buffer is full and has to discard arriving packets from many TCP connections, they will all enter the slow-start phase. This significantly reduces the load through the buffer, leading to a period of under-utilization. Then all those TCP connections will come out of slow-start at about the same time, leading to a substantial increase in traffic and causing congestion in the buffer. More packets are discarded, and the cycle repeats – this is called 'global synchronization'.

Random early detection (RED) is a packet discard mechanism that anticipates congestion by discarding packets *probabilistically* before the buffer becomes full [16.1]. It does this by monitoring the average queue size, and discarding packets with increasing probability when this average is above a configurable threshold, θ_{min}. Thus in the early stages of congestion, only a few TCP connections are affected, and this may be sufficient to reduce the load and avoid any further increase in congestion. If the average queue size continues to increase, then packets are discarded with increasing probability, and so more TCP connections are affected. Once the average queue size exceeds an upper threshold, θ_{max}, all arriving packets are discarded.

Why is the *average* queue size used – why not use the actual queue size (as with partial buffer sharing (PBS) in ATM)? Well, in ATM we have two different levels of space priority, and PBS is an algorithm for providing two distinct levels of cell loss probability. The aim of RED is to avoid congestion, not to differentiate between priority levels and provide different loss probability targets. If actual queue sizes are used, then the scheme becomes sensitive to transient congestion – short-lived bursts which don't need to be avoided, but just require the temporary storage space of a large buffer. By using average queue size, these short-lived bursts are filtered out. Of course, the bursts will increase the average temporarily, but this takes some time to feed through and, if it is not sustained, the average will remain below the threshold.

The average is calculated using an exponentially weighted moving average (EWMA) of queue sizes. At each arrival, i, the average queue size, q_i, is updated by applying a weight, w, to the current queue size, k_i.

$$q_i = w \cdot k_i + (1 - w) \cdot q_{i-1}$$

How quickly q_i responds to bursts can be adjusted by setting the weight, w. In [16.1] a value of 0.002 is used for many of the simulation scenarios, and a value greater than or equal to 0.001 is recommended to ensure adequate calculation of the average queue size.

Let's take a look at how the EWMA varies for a sample set of packet arrivals. In Figure 16.1 we have a Poisson arrival process of packets, at a load of 90% of the server capacity, over a period of 5000 time units. The thin grey line shows the actual queue state, and the thicker black line shows the average queue size calculated using the EWMA formula with $w = 0.002$. Figure 16.2 shows the same trace with a value of 0.01 for the weight, w. It is clear that the latter setting is not filtering out much of the transient behaviour in the queue.

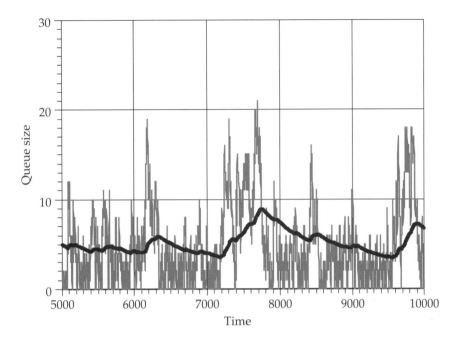

Figure 16.1. Sample Trace of Actual Queue Size (Grey) and EWMA (Black) with $w = 0.002$

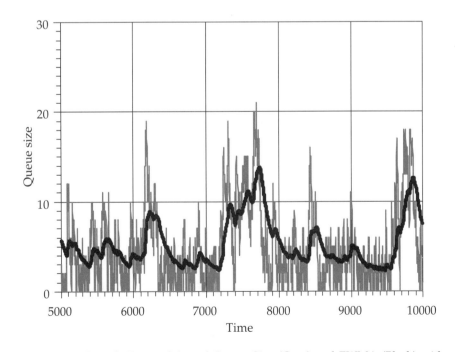

Figure 16.2. Sample Trace of Actual Queue Size (Grey) and EWMA (Black) with $w = 0.01$

Configuring the values of the thresholds, θ_{min} and θ_{max}, depends on the target queue size, and hence system load, required. In [16.1] a rule of thumb is given to set $\theta_{max} > 2\theta_{min}$ in order to avoid the synchronization problems mentioned earlier, but no specific guidance is given on setting θ_{min}. Obviously if there is not much difference between the thresholds, then the mechanism cannot provide sufficient advance warning of potential congestion, and it soon gets into a state where it drops all arriving packets. Also, if the thresholds are set too low, this will constrain the normal operation of the buffer, and lead to under-utilization. So, are there any useful indicators?

From the packet queueing analysis in the previous two chapters, we know that in general the queue state probabilities can be expressed as

$$p(k) = (1 - d_r) \cdot (d_r)^k$$

where d_r is the decay rate, k is the queue size and $p(k)$ is the queue state probability. The mean queue size can be found from this expression, as follows:

$$q = \sum_{k=1}^{\infty} k \cdot p(k) = (1 - d_r) \cdot \sum_{k=1}^{\infty} k \cdot d_r^k$$

Multiplying both sides by the decay rate gives

$$d_r \cdot q = (1 - d_r) \cdot \sum_{k=2}^{\infty} (k - 1) \cdot d_r^k$$

If we now subtract this equation from the previous one, we obtain

$$(1 - d_r) \cdot q = (1 - d_r) \cdot \sum_{k=1}^{\infty} d_r^k$$

$$q = \sum_{k=1}^{\infty} d_r^k$$

Multiplying both sides by the decay rate, again, gives

$$d_r \cdot q = \sum_{k=2}^{\infty} d_r^k$$

And, as before, we now subtract this equation from the previous one to obtain

$$(1 - d_r) \cdot q = d_r$$

$$q = \frac{d_r}{1 - d_r}$$

For the example shown in Figures 16.1 and 16.2, assuming a fixed packet size (i.e. the M/D/1 queue model) and using the GAPP formula with a load of 0.9 gives a decay rate of

$$d_r = \left. \frac{\lambda \cdot e^{\lambda} - e^{\lambda} - \lambda^2 + \lambda + e^{-\lambda}}{\lambda - 1 + e^{-\lambda}} \right|_{\lambda=0.9} = 0.817$$

and a mean queue size of

$$q = \frac{0.817}{1 - 0.817} = 4.478$$

which is towards the lower end of the values shown on the EWMA traces.

Figure 16.3 gives some useful indicators to aid the configuration of the thresholds, θ_{min} and θ_{max}. These curves are for both the mean queue size against decay rate, and for various levels of probability of exceeding a threshold queue size. Recall that the latter is given by

$$\Pr\{\text{queue size} > k\} = Q(k) = d_r^{k+1}$$

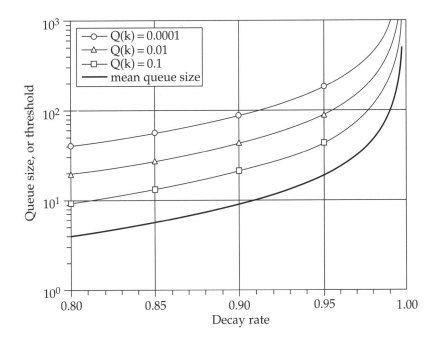

Figure 16.3. Design Guide to Aid Configuration of Thresholds, Given Required Decay Rate

So, to find the threshold k, given a specified probability, we just take logs of both sides and rearrange thus:

$$\text{threshold} = \frac{\log(\Pr\{\text{threshold exceeded}\})}{\log(d_r)} - 1$$

Note that this defines a threshold in terms of the probability that the *actual* queue size exceeds the threshold, not the probability that the EWMA queue size exceeds the threshold. But it does indicate how the queue behaviour deviates from the mean size in heavily loaded queues.

But what if we want to be sure that the mechanism can cope with a certain level of bursty traffic, without initiating packet discard? Recall the scenario in Chapter 15 for multiplexing an aggregate of packet flows. There, we found that although the queue behaviour did not go into the excess-rate ON state very often, when it did, the bursts could have a substantial impact on the queue (producing a decay rate of 0.964 72). It is thus the *conditional* behaviour of the queueing above the long-term average which needs to be taken into account. In this particular case, the decay rate of 0.964 72 has a mean queue size of

$$q = \frac{0.964\,72}{1 - 0.964\,72} = 27.345 \text{ packets}$$

The long-term average load for the scenario is

$$\lambda = \frac{5845}{7302.5} = 0.8$$

If we consider this as a Poisson stream of arrivals, and thus neglect the bursty characteristics, we obtain a decay rate of

$$d_r = \left. \frac{\lambda \cdot e^{\lambda} - e^{\lambda} - \lambda^2 + \lambda + e^{-\lambda}}{\lambda - 1 + e^{-\lambda}} \right|_{\lambda=0.8} = 0.659$$

and a long-term average queue size of

$$q = \frac{0.659}{1 - 0.659} = 1.933 \text{ packets}$$

It is clear, then, that the conditional behaviour of bursty traffic dominates the shorter-term average queue size. This is additional to the longer-term average, and so the sum of these two averages, i.e. 29.3, gives us a good indicator for the minimum setting of the threshold, θ_{min}.

VIRTUAL BUFFERS AND SCHEDULING ALGORITHMS

The disadvantage of the FIFO buffer is that all the traffic has to share the buffer space and server capacity, and this can lead to problems such as global synchronization as we saw in the previous section. The principle behind the RED algorithm is that it applies the 'brakes' gradually – initially affecting only a few end-to-end connections. Another approach is to partition the buffer space into virtual buffers, and use a scheduling mechanism to divide up the server capacity between them. Whether the virtual buffers are for individual flows, aggregates, or classes of flows, the partitioning enables the delay and loss characteristics of the individual virtual buffers to be tailored to specific requirements. This helps to contain any unwanted congestion behaviour, rather than allowing it to have an impact on all traffic passing through a FIFO output port. Of course, the two approaches are complementary – if more than one flow shares a virtual buffer, then applying the RED algorithm *just to that virtual buffer* can avoid congestion for those particular packet flows.

Precedence queueing

There are a variety of different scheduling algorithms. In Chapter 13, we looked at time priorities, also called 'head-of-line' (HOL) priorities, or precedence queueing in IP. This is a static scheme: each arriving packet has a fixed, previously defined, priority level that it keeps for the whole of its journey across the network. In IPv4, the Type of Service (TOS) field can be used to determine the priority level, and in IPv6 the equivalent field is called the Priority Field. The scheduling operates as follows (see Figure 16.4): packets of priority 2 will be served only if there are no packets

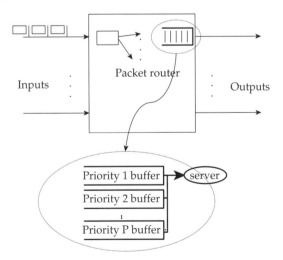

Figure 16.4.   HOL Priorities, or Precedence Queueing, in IP

of priorities 1; packets of priority 3 will be served only if there are no packets of priorities 1 and 2, etc. Any such system, when implemented in practice, will have to predefine P, the number of different priority classes.

From the point of view of the queueing behaviour, we can state that, in general, the highest-priority traffic sees the full server capacity, and each next highest level sees what is left over, etc. In a system with variable packet lengths, the analysis is more complicated if the lower-priority traffic streams tend to have larger packet sizes. Suppose a priority-2 packet of 1000 octets has just entered service (because the priority-1 virtual buffer was empty), but a short 40-octet priority-1 packet turns up immediately after this event. This high-priority packet must now wait until the lower-priority packet completes service – during which time as many as 25 such short packets could have been served.

Weighted fair queueing

The problem with precedence queueing is that, if the high-priority loading on the output port is too high, low-priority traffic can be indefinitely postponed. This is not a problem in ATM because the traffic control framework requires resources to be reserved and assessed in terms of the end-to-end quality of service provided. In a best-effort IP environment the build-up of a low-priority queue will not affect the transfer of high-priority packets, and therefore will not cause their end-to-end transport-layer protocols to adjust.

An alternative is round robin scheduling. Here, the scheduler looks at each virtual buffer in turn, serving one packet from each, and passing over any empty virtual buffers. This ensures that all virtual buffers get some share of the server capacity, and that no capacity is wasted. However, short packets are penalized – the end-to-end connections which have longer packets get a greater proportion of the server capacity because it is shared out according to the number of packets.

Weighted fair queueing (WFQ) shares out the capacity by assigning weights to the service of the different virtual buffers. If these weights are set according to the token rate in the token bucket specifications for the flows, or flow aggregates, and resource reservation ensures that the sum of the token rates does not exceed the service capacity, then WFQ scheduling effectively enables each virtual buffer to be treated independently with a service rate equal to the token rate.

If we combine WFQ with per-flow queueing (Figure 16.5), then the buffer space and server capacity can be tailored according to the delay and loss requirements of each flow. This is optimal in a traffic control sense because it ensures that badly behaved flows do not cause excessive delay or loss among well-behaved flows, and hence avoids the global synchronization problems. However, it is non-optimal in the overall loss

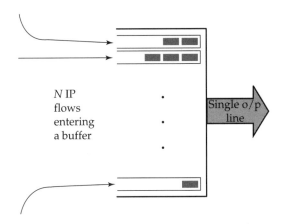

Figure 16.5. Per-flow Queueing, with WFQ Scheduling

sense: it makes far worse use of the available space than would, for example, complete sharing of a buffer. This can be easily seen when you realize that a single flow's virtual buffer can overflow, so causing loss, even when there is still plenty of space available in the rest of the buffer.

Each virtual buffer can be treated independently for performance analysis, so any of the previous approaches covered in this book can be re-used. If we have per-flow queueing, then the input traffic is just a single source. With a variable-rate flow, the peak rate, mean rate and burst length can be used to characterize a single ON–OFF source for queueing analysis. If we have per-class queueing, then whatever is appropriate from the M/D/1, M/G/1 or multiple ON–OFF burst-scale analyses can be applied.

BUFFER SPACE PARTITIONING

We have covered a number of techniques for calculating the decay rate, and hence loss probability, at a buffer, given certain traffic characteristics. In general, the loss probability can be expressed in terms of the decay rate, d_r, and buffer size, X, thus:

$$\text{loss probability} \approx \Pr\{\text{queue size} > X\} = Q(X) = d_r^{X+1}$$

This general form can easily be rearranged to give a dimensioning formula for the buffer size:

$$X \approx \frac{\log(\text{loss probability})}{\log(d_r)} - 1$$

For realistically sized buffers, one packet space will make little difference, so we can simplify this equation further to give

$$X \approx \frac{\log(\text{loss probability})}{\log(d_r)}$$

But many manufacturers of switches and routers provide a certain amount of buffer space, X, at each output port, which can be partitioned between the virtual buffers according to the requirements of the different traffic classes/aggregates. The virtual buffer partitions are configurable under software control, and hence must be set by the network operator in a way that is consistent with the required loss probability (LP) for each class.

Let's take an example. Recall the scenario for Figure 14.10. There were three different traffic aggregates, each comprising a certain proportion of long and short packets, and with a mean packet length of 500 octets. The various parameters and their values are given in Table 16.1.

Suppose each aggregate flow is assigned a virtual buffer and is served at one third of the capacity of the output port, as shown in Figure 16.6. If we want all the loss probabilities to be the same, how do we partition the available buffer space of 200 packets (i.e. 100 000 octets)? We require

$$LP \approx dr_1{}^{X_1} = dr_2{}^{X_2} = dr_3{}^{X_3}$$

given that

$$X_1 + X_2 + X_3 = X = 200 \text{ packets}$$

By taking logs, and rearranging, we have

$$X_1 \cdot \log(dr_1) = X_2 \cdot \log(dr_2) = X_3 \cdot \log(dr_3)$$

Table 16.1.   Parameter Values for Bi-modal Traffic Aggregates

Parameter	Bi-modal 540	Bi-modal 960	Bi-modal 2340
Short packets (octets)	40	40	40
Long packets (octets)	540	960	2340
Ratio of long to short, n	13.5	24	58.5
Proportion of short packets, p_s	0.08	0.5	0.8
Packet arrival rate, λ	0.064	0.064	0.064
E[a]	0.8	0.8	0.8
a(0)	0.4628	0.57 662	0.75 514
a(1)	0.33 982	0.19 532	0.06 574
Decay rate, d_r	0.67 541	0.78 997	0.91 454

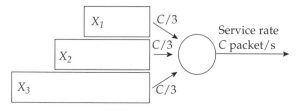

Figure 16.6.   Example of Buffer Space Partitioning

and so

$$X_2 = \frac{X_1 \cdot \log(dr_1)}{\log(dr_2)}$$

$$X_3 = \frac{X_1 \cdot \log(dr_1)}{\log(dr_3)}$$

Summing up both sides of these equations gives

$$X_2 + X_3 = X - X_1 = \frac{X_1 \cdot \log(dr_1)}{\log(dr_2)} + \frac{X_1 \cdot \log(dr_1)}{\log(dr_3)}$$

To make it a generic formula, we have

$$X - X_1 = \frac{X_1 \cdot \log(dr_1)}{\log(dr_1)} + \left\{ \frac{X_1 \cdot \log(dr_1)}{\log(dr_2)} + \frac{X_1 \cdot \log(dr_1)}{\log(dr_3)} \right\} - \frac{X_1 \cdot \log(dr_1)}{\log(dr_1)}$$

$$X - X_1 = \sum_{j=1}^{3} \left(\frac{X_1 \cdot \log(dr_1)}{\log(dr_j)} \right) - X_1$$

$$X = X_1 \cdot \log(dr_1) \cdot \sum_{j=1}^{3} \left(\frac{1}{\log(dr_j)} \right)$$

So we can write

$$X_i = \frac{X}{\log(dr_i) \cdot \displaystyle\sum_{j=1}^{3} \left(\frac{1}{\log(dr_j)} \right)}$$

In our example, we have

$$X = 200 \text{ packets}$$

$$dr_1 = 0.675\,41$$

$$dr_2 = 0.789\,97$$

$$dr_3 = 0.914\,54$$

By applying our partitioning formula, we obtain (to the nearest whole packet)

$$X_1 = 28 \text{ packets}$$

$$X_2 = 47 \text{ packets}$$

$$X_3 = 125 \text{ packets}$$

This gives loss probabilities for each of the virtual buffers of approximately 1.5×10^{-5}.

If we want to achieve different loss probabilities for each of the traffic classes, we can introduce a scaling factor, S_i, associated with each traffic class. For example, we may require the loss probabilities in our example to be related as

$$10\,000 \cdot LP_1 = 100 \cdot LP_2 = 1 \cdot LP_3$$

i.e.

$$S_1 = 10\,000$$

$$S_2 = 100$$

$$S_3 = 1$$

Let's generalize to any number, V, of virtual buffers, and modify the previous approach by allowing for any scaling of loss probability.

$$S_1 \cdot dr_1{}^{X_1} = S_2 \cdot dr_2{}^{X_2} = \cdots = S_j \cdot dr_j{}^{X_j} = \cdots = S_V \cdot dr_V{}^{X_V}$$

$$X = \sum_{j=1}^{V} X_j$$

By taking logs, and rearranging, we have

$$\log(S_1) + X_1 \cdot \log(dr_1) = \log(S_2) + X_2 \cdot \log(dr_2)$$

$$= \cdots = \log(S_V) + X_V \cdot \log(dr_V)$$

and so

$$X_2 = \frac{\log(S_1) + X_1 \cdot \log(dr_1) - \log(S_2)}{\log(dr_2)}$$

$$\vdots$$

$$X_V = \frac{\log(S_1) + X_1 \cdot \log(dr_1) - \log(S_V)}{\log(dr_V)}$$

After summing up both sides, and some rearrangement, we have the general formula

$$X_i = \frac{X + \sum_{j=1}^{V} \left(\dfrac{\log(S_j)}{\log(dr_j)} \right)}{\log(dr_i) \cdot \sum_{j=1}^{V} \left(\dfrac{1}{\log(dr_j)} \right)} - \frac{\log(S_i)}{\log(dr_i)}$$

This buffer partitioning formula can be used to evaluate the correct size of the partition allocated to any traffic class depending only on knowing the total space available, the decay rates and the desired scaling factors.

For our example with three virtual buffers, we have

$$X = 200 \text{ packets}$$
$$dr_1 = 0.675\,41$$
$$dr_2 = 0.789\,97$$
$$dr_3 = 0.914\,54$$

and

$$S_1 = 10\,000$$
$$S_2 = 100$$
$$S_3 = 1$$

By applying the general partitioning formula, we obtain (to the nearest whole packet)

$$X_1 = 46 \text{ packets}$$
$$X_2 = 56 \text{ packets}$$
$$X_3 = 98 \text{ packets}$$

This gives loss probabilities for each of the virtual buffers of

$$LP_1 = 1.446 \times 10^{-8}$$
$$LP_2 = 1.846 \times 10^{-6}$$
$$LP_3 = 1.577 \times 10^{-4}$$

SHARED BUFFER ANALYSIS

Earlier we noted that partitioning a buffer is non-optimal in the overall loss sense. Indeed, if buffer space is shared between multiple output ports, much better use can be made of the resource (see Figure 16.7). But can we quantify this improvement? The conventional approach is to take the convolution of the individual state probability distributions, in order to combine all the possibilities of having a certain total number in the buffer. Assuming the same state probability distributions for each of the output buffers, and that the arrivals to each buffer are independent of each other, let

$$P_N(k) = \Pr\{\text{queue state for } N \text{ buffers sharing} = k\}$$

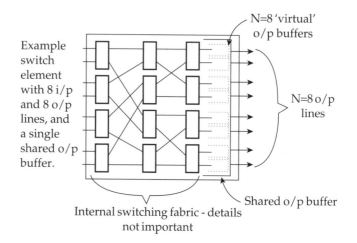

Figure 16.7.   Example of a Switch/Router with Output Ports Sharing Buffer Space

The autoconvolution for two buffers sharing is given by

$$P_2(k) = \sum_{j=0}^{k} P_1(j) \cdot P_1(k - j)$$

i.e. to find the probability that the shared buffer is in state k, find all the different ways in which the two individual queues can have k packets between them, and sum these probabilities.

The autoconvolution for N buffers sharing the buffer space can then be constructed recursively, i.e.

$$P_N(k) = \sum_{j=0}^{k} P_{N-1}(j) \cdot P_1(k - j)$$

and the loss probability estimated from

$$Q_N(k) = 1 - \sum_{j=0}^{k} P_N(j)$$

From the packet queueing analysis in the previous two chapters, we know that in general the state probabilities for an individual queue can be expressed as

$$p(k) = (1 - d_r) \cdot (d_r)^k$$

where d_r is the decay rate, k is the queue size, and $p(k)$ is the individual queue state probability. The autoconvolution of this geometric distribution is given by a negative binomial; thus

$$P_N(k) = {}^{k+N-1}C_{N-1} \cdot (d_r)^k \cdot (1 - d_r)^N$$

where k is the size of the combined queues *in the shared buffer*.

The probability that the shared buffer overflows is expressed as

$$Q_N(k-1) = 1 - \sum_{j=0}^{k-1} P_N(j) = \sum_{j=k}^{\infty} P_N(j)$$

To avoid having to do summations, a geometric approximation can be applied to the tail of the negative binomial, i.e.

$$Q_N(k-1) = \sum_{j=k}^{\infty} P_N(j) \approx \sum_{j=k}^{\infty} P_N(k) \cdot q^{j-k}$$

Note that this is, in essence, the same approach that we used previously in Chapter 14 for the Geometrically Approximated Poisson Process. However, we cannot parameterize it via the mean of the excess-rate batch size – instead we estimate the geometric parameter, q, from the ratio of successive queue state probabilities:

$$q = \frac{P_N(k+1)}{P_N(k)} = \frac{{}^{k+N}C_{N-1} \cdot (d_r)^{k+1} \cdot (1-d_r)^N}{{}^{k+N-1}C_{N-1} \cdot (d_r)^k \cdot (1-d_r)^N}$$

which, once the combinations have been expanded, reduces to

$$q = \frac{(k+N) \cdot d_r}{k+1}$$

For any practical arrangement in IP packet queueing, the buffer capacity will be large compared to the number of output ports sharing; so

$$q \approx d_r \quad \text{for } k \gg N$$

So, applying the geometric approximation, we have

$$Q_N(k-1) \approx P_N(k) \cdot \frac{1}{1-q}$$

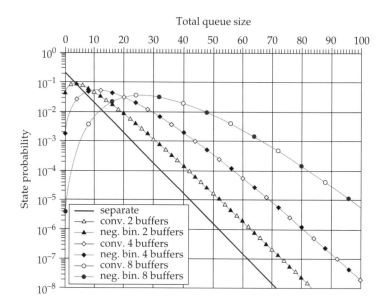

$$p(k, dr) := (1 - dr) \cdot dr^k$$

$$\text{Autoconv}(N, P) := \begin{vmatrix} P & \text{if} \quad N < 2 \\ \text{otherwise} \\ \quad \begin{vmatrix} Q \leftarrow p \\ \max k \leftarrow \text{last}(P) \\ i \leftarrow 1 \\ \text{while } i < N \\ \quad \begin{vmatrix} \text{for } m \in 0..\text{maxk} \\ \quad \text{conv}_m \leftarrow \sum_{j=0}^{m} P_j \cdot Q_{m-j} \\ Q \leftarrow \text{conv} \\ i \leftarrow i + 1 \end{vmatrix} \\ Q \end{vmatrix} \end{vmatrix}$$

$$\text{NegBinomial}(N, k, dr) := \text{combin}(k + N - 1, N - 1) \cdot dr^k \cdot (1 - dr)^N$$

$$k := 0.. \quad 100$$

$$dr := 0.78997$$

$$\text{Psingle}_k := p(k, dr)$$

$$x_k := k$$

$$y1 := \text{Psingle}$$

$$y2 := \text{Autoconv}(2, \text{Psingle})$$

$$y3_k := \text{NegBinomial}(2, k, dr)$$

$$y4 := \text{Autoconv}(4, \text{Psingle})$$

$$y5_k := \text{NegBinomial}(4, k, dr)$$

$$y6 := \text{Autoconv}(8, \text{Psingle})$$

$$y7_k := \text{NegBinomial}(8, k, dr)$$

Figure 16.8. State Probabilities for Shared Buffers, and Mathcad Code to Generate (x, y) Values for Plotting the Graph

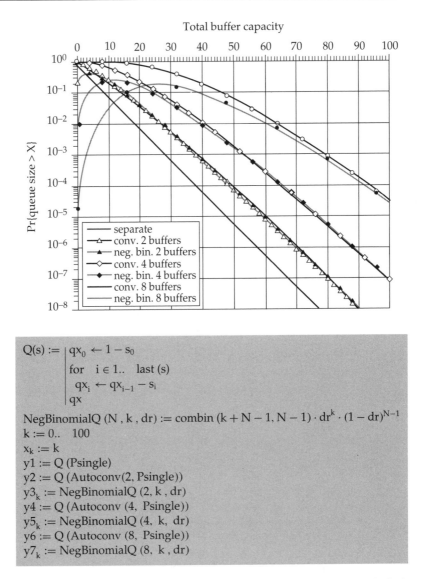

Figure 16.9. Overflow Probabilities for Shared Buffers, and Mathcad Code to Generate (x, y) Values for Plotting the Graph

which, after substituting for $P_N(k)$ and q, gives

$$Q_N(k-1) \approx {}^{k+N-1}C_{N-1} \cdot (d_r)^k \cdot (1-d_r)^{N-1}$$

Applying Stirling's approximation, i.e.

$$N! = \frac{N^N \cdot e^{-N}}{\sqrt{2 \cdot \pi \cdot N}}$$

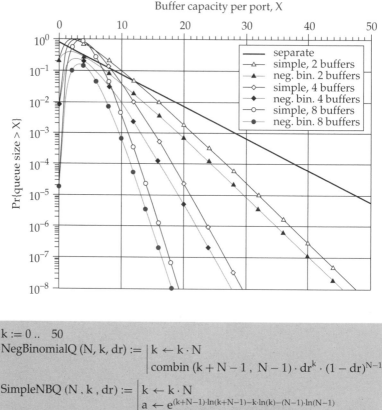

Buffer capacity per port, X

$$k := 0 .. \ 50$$

$$\text{NegBinomialQ}\,(N, k, dr) := \begin{vmatrix} k \leftarrow k \cdot N \\ \text{combin}\,(k + N - 1,\ N - 1) \cdot dr^{k} \cdot (1 - dr)^{N-1} \end{vmatrix}$$

$$\text{SimpleNBQ}\,(N, k, dr) := \begin{vmatrix} k \leftarrow k \cdot N \\ a \leftarrow e^{(k+N-1)\cdot \ln(k+N-1)-k\cdot \ln(k)-(N-1)\cdot \ln(N-1)} \\ b \leftarrow e^{k\cdot \ln(dr)+(N-1)\cdot \ln(1-dr)} \\ a \cdot b \end{vmatrix}$$

$$x_k := k$$
$$y1 := Q\,(\text{Psingle})$$
$$y2_k := \text{SimpleNBQ}\,(2, k, dr)$$
$$y3_k := \text{NegBinomialQ}\,(2, k, dr)$$
$$y4_k := \text{SimpleNBQ}\,(4, k, dr)$$
$$y5_k := \text{NegBinomialQ}\,(4, k, dr)$$
$$y6_k := \text{SimpleNBQ}\,(8, k, dr)$$
$$y7_k := \text{NegBinomialQ}\,(8, k, dr)$$

Figure 16.10. Comparison Showing the Benefits of Sharing Buffer Space – Overflow Probability vs. Buffer Capacity per Output Port

gives

$$Q_N(k - 1) \approx \frac{(k + N - 1)^{k+N-1}}{k^k \cdot (N - 1)^{N-1}} \cdot (d_r)^k \cdot (1 - d_r)^{N-1}$$

which has the distinct advantage of not requiring the user to evaluate large factorials. Applying logarithms ensures that all the powers can be evaluated without number overflow.

$$Q_N(k-1) \approx e^{\{(k+N-1)\cdot\ln(k+N-1)-k\cdot\ln(k)-(N-1)\cdot\ln(N-1)+k\cdot\ln(d_r)+(N-1)\cdot\ln(1-d_r)\}}$$

Thus we have a simplified negative binomial expression for the overflow probability in terms of the decay rates in (conceptually) separate queues [16.2].

Let's now suppose we have a number of output ports sharing buffer space, and each output port is loaded to 80% of its server capacity with a bi-modal traffic aggregate (e.g. column 2 in Table 16.1 – bi-modal 960). The decay rate, assuming no buffer sharing, is 0.789 97. Figure 16.8 compares the state probabilities based on exact convolution with those based on the negative binomial expression, clearly showing the precision of the latter approach. Figure 16.9 compares the overflow probabilities based on exact convolution with those based on the negative binomial expression (using ^nC_r). There is an obvious discrepancy for very small shared buffers, but above this the comparison is very close indeed.

If the loss probability requirement is 10^{-4} then having separate buffers means that
$$38 \times 500 = 19\,000 \text{ octets}$$

are required per output port, whereas if the 8 output ports share buffer space then only
$$(94 \times 500)/8 = 5875 \text{ octets}$$

are required per output port (recall that the average packet size is 500 octets in this example).

Figure 16.10 shows this very clearly by plotting the overflow probability against the buffer capacity *per output port*. In this case we compare the negative binomial expression (using ^nC_r) with the simplified negative binomial expression (using logarithms).

17 Self-similar Traffic

play it again, Sam

SELF-SIMILARITY AND LONG-RANGE-DEPENDENT TRAFFIC

The queueing models and solutions we have presented, developed and applied in this book are very useful and have wide applicability. However, one of the most significant recent findings for the design and performance evaluation of networks has been the discovery of self-similarity and long-range dependence (LRD) in a variety of traffic types [17.1]. Why is it significant? Well, the essence of self-similarity is that a time-varying process behaves in a similar way over all time scales. The observations made on a variety of traffic types in different network technologies show *bursty* behaviour over *a wide range* of time scales. And, as we have seen in previous chapters, bursty behaviour has a much greater impact on finite resources.

Let's take a memoryless process first, and see how that scales with time. Figure 17.1 shows the results of simulating traffic for 10 000 seconds. The first 100 seconds of the arrival process are shown as a thin grey line, and here we see typical variable behaviour around a mean value of about 25 arrivals per second. The thick black line shows the process scaled by 100, i.e. the number of arrivals is averaged every 100 seconds and so the 100 scaled time units cover the full 100 00 seconds of the simulation. This averaging clearly shows a reduction in the variability of the process when viewed on the longer time scale – the mean value of 25 arrivals per second is evident. Figure 17.2 takes a self-similar process and plots it in the same way. In this case we can see the high variability of the process even after scaling.

However, it is not self-similarity which is the underlying phenomenon, but rather it is the presence of many basic communications processes which have heavy-tailed sojourn-time distributions. In these distributions, the tail probabilities decay as a power law, rather than exponentially.

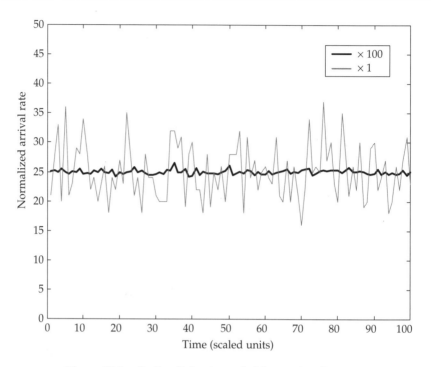

Figure 17.1.   Scaling Behaviour of a Memoryless Process

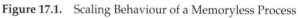

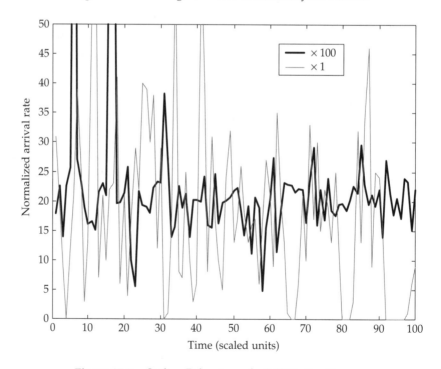

Figure 17.2.   Scaling Behaviour of a Self-Similar Process

Another way of expressing this is that the process has long-term (slowly decaying) correlations. So, an individual communications process with heavy-tailed sojourn times exhibits long-range dependence. And the aggregation of LRD sources produces a traffic stream with self-similar characteristics.

So, how do we model and analyse the impact of this traffic? There have been claims that 'traditional' approaches to teletraffic modelling no longer apply. Much research effort has been, and is being, spent on developing new teletraffic models, such as Fractional Brownian Motion (FBM) processes (e.g. [17.2]) and non-linear chaotic maps (e.g. [17.3]). However, because of their mathematical complexity, assessing their impact on network resources is not a simple task, although good progress is being made.

In this chapter we take a different approach: with a little effort we can re-apply what we already know about traffic engineering usefully, and generate results for these new scenarios quickly. Indeed, this is in line with our approach throughout this book.

THE PARETO MODEL OF ACTIVITY

A distribution is heavy-tailed if

$$\Pr\{X > x\} = 1 - F(x) \approx \frac{1}{x^\alpha}$$

as $x \to \infty$, and noting that $\alpha > 0$ (usually α takes on values in the range $1 \to 2$). The Pareto distribution is one of the class of distributions that are 'heavy-tailed', and is defined as

$$\Pr\{X > x\} = \left(\frac{\delta}{x}\right)^\alpha$$

where δ is the parameter which specifies the minimum value that the distribution can take, i.e. $x \geqslant \delta$. For example, if $\delta = 25$, then $\Pr\{X > 25\} = 1$, i.e. X cannot be less than or equal to 25. For our purposes it is often convenient to set $\delta = 1$.

The cumulative distribution function is

$$F(x) = 1 - \left(\frac{\delta}{x}\right)^\alpha$$

and the probability density function is given by

$$f(x) = \frac{\alpha}{\delta} \cdot \left(\frac{\delta}{x}\right)^{\alpha+1}$$

The mean value of the Pareto distribution is

$$E[x] = \delta \cdot \frac{\alpha}{\alpha - 1}$$

Note that for this formula to be correct, $\alpha > 1$ is essential; otherwise the Pareto has an infinite mean.

Let's put some numbers in to get an idea of the effect of moving to heavy-tailed distributions. Assume that we have a queue with a time-slotted arrival process of packets or cells. The load is 0.5, and we have a batch arriving as a Bernoulli process, such that

$$Pr\{\text{there is a batch in a time slot}\} = 0.25$$

thus the mean number of arrivals in any batch is 2. We calculate the probability of having more than x arrivals in any time slot, in two cases: for an exponentially distributed batch size, and for a Pareto-distributed batch size. In the former case, we have

$$Pr\{\text{batch size} > x\} = e^{-\frac{x}{2}}$$

so

$$Pr\{> 10 \text{ arrivals in any time slot}\} = Pr\{\text{batch size} > 10\}$$
$$\times Pr\{\text{there is a batch in a time slot}\}$$
$$= e^{-\frac{10}{2}} \times 0.25 = 0.001\,684$$

In the latter case, we have (with $\delta = 1$)

$$E[x] = 1 \cdot \frac{\alpha}{\alpha - 1} = 2$$

so

$$\alpha = \frac{E[x]}{E[x] - 1} = 2$$

hence

$$Pr\{\text{batch size} > x\} = \left(\frac{1}{x}\right)^2$$

giving

$$Pr\{>10 \text{ arrivals in any time slot}\} = \left(\frac{1}{10}\right)^2 \times 0.25 = 0.0025$$

Thus for a batch size of greater than 10 arrivals there is not that much difference between the two distributions – the probability is of the same order of magnitude. However, if we try again for more than 100 arrivals we obtain

$$\text{Pr}\{>100 \text{ arrivals in any time slot}\} = e^{-\frac{100}{2}} \times 0.25 = 4.822 \times 10^{-23}$$

in the exponential case, and

$$\text{Pr}\{>100 \text{ arrivals in any time slot}\} = \left(\frac{1}{100}\right)^2 \times 0.25 = 2.5 \times 10^{-5}$$

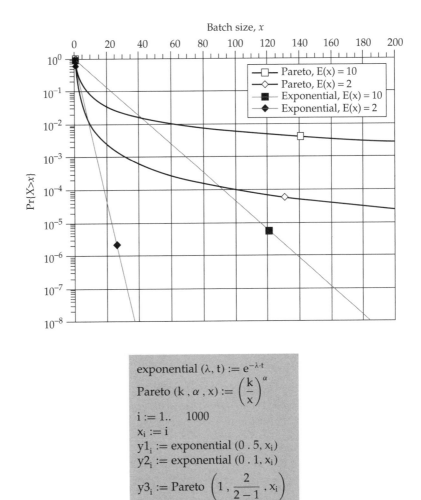

Figure 17.3. Comparison of Exponential and Pareto Distributions, and the Mathcad Code to Generate (x, y) Values for Plotting the Graph

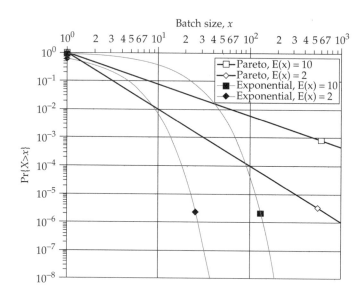

Figure 17.4. Comparison of Exponential and Pareto Distributions, with Logarithmic Scale for x

in the Pareto case. This is a *significant* difference, and clearly illustrates the problems associated with highly variable traffic, i.e. non-negligible probabilities for large batch sizes, or long sojourn times.

Figure 17.3 compares the exponential and Pareto distributions for two different mean batch sizes, plotting x on a linear scale. For the exponential distribution (which we have used extensively for sojourn times in state-based models) the logarithm of the probability falls away linearly with increasing x. But for the Pareto the distribution 'bends back' so that much longer values have much more significant probability values than they would otherwise. In fact we can see, in Figure 17.4, that when both axes have a logarithmic scale, there is a straight-line relationship for the Pareto.

We can see from these figures that the Pareto distribution has increasing, not constant, decay rate. This is *very* important for our analysis; for example, as the ON period continues, the probability of the ON period coming to an end diminishes. This is completely different from the exponential model, and the effect on buffer content is predictably dramatic.

IMPACT OF LRD TRAFFIC ON QUEUEING BEHAVIOUR

In previous queueing analysis we have been able to use memoryless distributions such as the exponential or geometric, in the traffic models, resulting in constant decay rates for the queueing behaviour. The effect of using a Pareto distribution is that, as the buffer fill becomes very large, the decay rate of the buffer-state probabilities tends to 1. This has an important practical outcome: above a certain level, there is no practical value in adding more buffer space to that already available. This is clearly both important and very different from those queueing systems we have already studied. The queue with Pareto-distributed input is then one of those examples (referred to previously in Chapter 14) which are not covered by the rule of asymptotically constant decay rates – except that it will always eventually be the case that the decay rate tends to 1!

The Geo/Pareto/1 queue

In order to explore the effects of introducing heavy-tailed distributions into the analysis, we can re-use the queueing analysis developed in Chapter 7. Let's assume a queue model in which batches of packets arrive at random, i.e. as a Bernoulli process, and the number of packets in a batch is Pareto-distributed. The Bernoulli process has a basic time unit (e.g. the time to serve an average-length packet), and a probability, q, that a batch arrives during the time unit. This is illustrated in Figure 17.5.

In order to use the queueing analysis from Chapter 7, we need to calculate the batch arrivals distribution. The probability that there are k arrivals in any time unit is denoted $a(k)$. Thus we write

$$a(0) = 1 - q$$

$$a(1) = q \cdot b(1)$$

$$a(2) = q \cdot b(2)$$

$$\vdots$$

$$a(k) = q \cdot b(k)$$

where $b(k)$ is the probability that an arriving batch has k packets. Note that this is a discrete distribution, whereas the Pareto, as defined earlier,

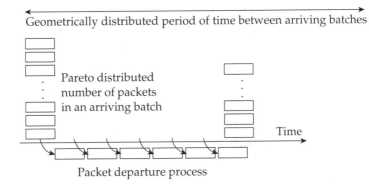

Geometrically distributed period of time between arriving batches

Pareto distributed
number of packets
in an arriving batch

Time

Packet departure process

Figure 17.5.   Model of Arriving Batches of Packets

is a continuous distribution. We use the cumulative form

$$F(x) = 1 - \left(\frac{1}{x}\right)^\alpha$$

to compute a discrete version of the Pareto distribution. In order to calculate $b(k)$, we use the interval $[k - 0.5, k + 0.5]$ on the continuous

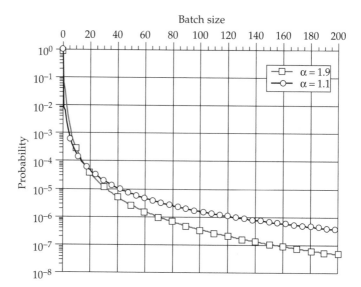

$$\text{BatchPareto}(q, k, \alpha) := \begin{vmatrix} 1 - q & \text{if } k = 0 \\[2mm] \left[1 - \left(\dfrac{1}{1.5}\right)^\alpha\right] \cdot q & \text{if } k = 1 \\[3mm] q \cdot \left[\left(\dfrac{1}{k - 0.5}\right)^\alpha - \left(\dfrac{1}{k + 0.5}\right)^\alpha\right] & \text{otherwise} \end{vmatrix}$$

$\text{maxX} := 1000$
$k := 0 .. \ \text{maxX}$
$l := 0 .. \ 1$
$\alpha := \begin{pmatrix} 1.9 \\ 1.1 \end{pmatrix}$
$B_l := \dfrac{\alpha_l}{\alpha_l - 1}$
$\rho := 0.25$
$q_l := \dfrac{\rho}{B_l}$
$x_k := k$
$y1_k := \text{BatchPareto}(q_0, k, \alpha_0)$
$y2_k := \text{BatchPareto}(q_1, k, \alpha_1)$

Figure 17.6. Discrete Version of Batch Pareto Input Distributions

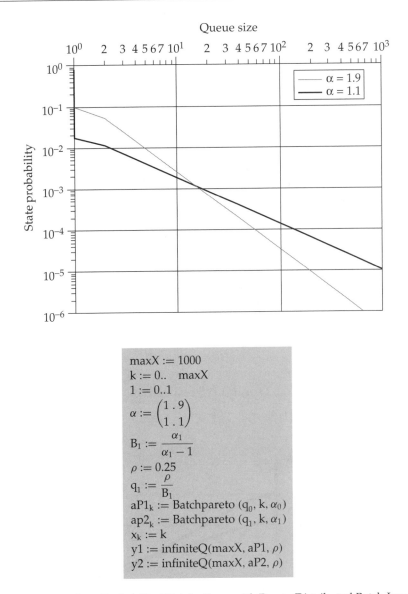

Figure 17.7. State Probability Distributions with Pareto Distributed Batch Input

distribution, i.e.

$$b(x) = F(x + 0.5) - F(x - 0.5) = \left(\frac{1}{x - 0.5}\right)^\alpha - \left(\frac{1}{x + 0.5}\right)^\alpha$$

Note that $F(1) = 0$, i.e. the probability that an arriving batch is less than or (exactly) equal to 1 packet is zero. Remember this is for a continuous distribution; so, for the discrete case of a batch size of one packet, we have

$$b(1) = F(1.5) - F(1) = 1 - \left(\frac{1}{1.5}\right)^{\alpha}$$

So, $b(k)$ is the conditional probability distribution for the number of packets arriving in a time unit, i.e. *given that there is a batch*; and $a(k)$ is the unconditional probability distribution for the number of packets arriving in a time unit – i.e. *whether there is an arriving batch or not.*

Intuitively, we can see that the probability there are no arrivals at all will probably be the biggest single value in the distribution – most of the time there will be zero arrivals, but when packets do arrive – *watch out* – because there are likely to be a lot of them! Figure 17.6 shows some example distributions for batch Pareto input, with $\alpha = 1.1$ and 1.9. The figure is plotted on a linear axis for the batch size, so that we can see the probability of no arrivals. Note that the mean batch sizes are 11 and 2.111 packets respectively. The mean number of packets per time unit is set to 0.25; thus the probability of there being a batch is

$$q = \frac{\rho}{B} = \frac{0.25}{B}$$

giving $q = 0.023$ and 0.118 respectively.

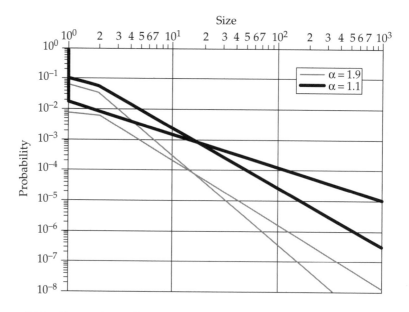

Figure 17.8. Comparison of Power-Law Decays for Arrival (Thin) and Queue-State (Thick) Probability Distributions

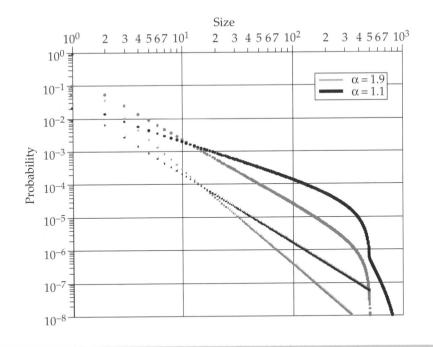

$$\text{BatchparetoTrunc}\,(q\,,k\,,\alpha\,,X) := \begin{vmatrix} 1 - q \text{ if } k = 0 \\[2mm] \dfrac{\left[1 - \left(\dfrac{1}{1\,.\,5}\right)^{\alpha}\right]}{1 - \left(\dfrac{1}{X+0\,.\,5}\right)^{\alpha}} \cdot q \quad \text{if } k = 1 \\[6mm] q \cdot \dfrac{\left[\left(\dfrac{1}{k-0\,.\,5}\right)^{\alpha} - \left(\dfrac{1}{k+0\,.\,5}\right)^{\alpha}\right]}{1 - \left(\dfrac{1}{X+0.5}\right)^{\alpha}} \quad \text{if } (k \leqslant X) \wedge (k > 1) \\[6mm] 0 \text{ if } k > X \end{vmatrix}$$

$\text{Xlimit} := 500$

$\text{aP1}_k := \text{BatchParetoTrunc}\,(q_0, k, \alpha_0, \text{Xlimit})$

$\text{aP2}_k := \text{BatchParetoTrunc}\,(q_1, k, \alpha_1, \text{Xlimit})$

$\rho\text{alt1} := \displaystyle\sum_k k.\text{aP1}_k$

$\rho\text{alt1} = 0.242$

$\rho\text{alt2} := \displaystyle\sum_k k.\text{aP2}_k$

$\rho\text{alt2} = 0.115$

$\text{y1} := \text{infiniteQ}\,(\text{maxX}, \text{aP1}, \rho\text{alt1})$

$\text{y2} := \text{infiniteQ}\,(\text{maxX}, \text{aP2}, \rho\text{alt1})$

Figure 17.9. Effect of Truncated Power-Law Decays for Arrival (Thin) and Queue-State (Thick) Probability Distributions

Now that we have prepared the arrival distribution, we can put this directly into the queueing analysis from Chapter 7. Figure 17.7 shows the resulting queue state probabilities for both $\alpha = 1.1$ and 1.9. Note that the queue-state probabilities have power-law decay similar to, but not the same as, the arrival distributions. This is illustrated in Figure 17.8, which shows the arrival probabilities as thin lines and the queue-state probabilities as thick lines.

From these results it appears that the advantage of having a large buffer is somewhat diminished by having to cope with LRD traffic: no buffer would seem to be large enough! However, in practice there is an upper limit to the time scales of correlated traffic activity. We can model this by truncating the Pareto distribution, and simply using the same approach to the queueing analysis.

Suppose X is the maximum number of packets in a batch. Our truncated, discrete version of the Pareto distribution now looks like

$$
b(x) = \begin{cases}
\left(1 - \left(\frac{1}{0.5}\right)^{\alpha}\right) \Big/ \left(1 - \left(\frac{1}{X+0.5}\right)^{\alpha}\right) & x = 1 \\
\left(\left(\frac{1}{x-0.5}\right)^{\alpha} - \left(\frac{1}{x+0.5}\right)^{\alpha}\right) \Big/ \left(1 - \left(\frac{1}{X+0.5}\right)^{\alpha}\right) & X \geqslant x > 1 \\
0 & x > X
\end{cases}
$$

Note that, because of the truncation, the probability density needs to be conditioned on what remains, i.e.

$$
1 - \left(\frac{1}{X+0.5}\right)^{\alpha}
$$

Figure 17.9 shows the result of applying this arrival distribution to the queueing analysis from Chapter 7. In this case we have the same values as before for α, i.e. $\alpha = 1.1$ and 1.9, and we set $X = 500$. The load is reduced because of the truncation to 0.115 and 0.242 respectively. The figure shows both the truncated arrival distributions and the resulting queue-state distributions. For the latter, it is clear that the power-law decay begins to change, even before the truncation limit, towards an exponential decay.

So, we can see that it is important to know the actual limit of the ON period activity in the presence of LRD traffic, because it has such a significant effect on the buffer size needed.

References

[1.1] Griffiths, J.M. (ed.), *ISDN Explained: Worldwide Network and Applications Technology*, John Wiley & Sons, ISBN 0 471 93480 1 (1992)

[1.2] Cuthbert, L.G. and Sapanel, J-C., *ATM: the Broadband Telecommunications Solution*, The Institution of Electrical Engineers, ISBN 0 85296 815 9 (1993)

[1.3] Thomas, S.A., *IPng and the TCP/IP Protocols: Implementing the Next Generation Internet*, John Wiley & Sons, ISBN 0 471 13088 5 (1996)

[3.1] Flood, J.E., *Telecommunications Switching, Traffic and Networks*, Prentice Hall, ISBN 0 130 33309 3 (1995)

[3.2] Bear, D., *Principles of Telecommunication Traffic Engineering*, Peter Peregrinus (IEE), ISBN 0 86341 108 8 (1988)

[5.1] Law, A.M. and Kelton, W.D., *Simulation Modelling and Analysis*, McGraw-Hill, ISBN 0 07 100803 9 (1991)

[5.2] Pitts, J.M., Cell-rate modelling for accelerated simulation of ATM at the burst level, *IEE Proceedings Communications*, **142**, 6, December 1995

[6.1] Cosmas, J.P., Petit, G., Lehnert, R., Blondia, C., Kontovassilis, K., and Cassals, O., A review of voice, data and video traffic models for ATM, *European Transactions on Telecommunications*, **5**, 2, March 1994

[7.1] Pattavina, A., *Switching Theory: Architecture and Performance in Broadband ATM Networks*, John Wiley & Sons ISBN 0 471 96338 0 (1998)

[8.1] Roberts, J.W. and Virtamo, J.T., The superposition of periodic cell arrival processes in an ATM multiplexer, *IEEE Trans. Commun.*, **39**, 2, pp. 298–303

[8.2] Norros, I., Roberts, J.W., Simonian, A. and Virtamo, J.T., The superposition of variable bit rate sources in an ATM multiplexer, *IEEE JSAC*, **9**, 3, April 1991, pp. 378–387

[8.3] Schormans, J.A., Pitts, J.M., Clements, B.R. and Scharf, E.M., Approximation to M/D/1 for ATM CAC, buffer dimensioning and cell loss performance, *Electronics Letters*, **32**, 3, 1996, pp. 164–165

[9.1] Onvural, R., *Asynchronous Transfer Mode Networks: Performance Issues*, Artech House (1995)

[9.2] Schormans, J.A., Pitts, J.M. and Cuthbert, L.G., Exact fluid-flow analysis of single on/off source feeding an ATM buffer, *Electronics Letters*, **30**, 14, 7 July 1994, pp. 1116–1117

[9.3] Lindberger, K., Analytical methods for the traffical problems with statistical multiplexing in ATM networks, 13th International Teletraffic Congress, Copenhagen 1991, **14**: Teletraffic and datatraffic in a period of change

[10.1] ITU Recommendation I.371 TRAFFIC CONTROL AND CONGESTION CONTROL IN B-ISDN, August 1996

[10.2] ATM Forum AF-TM-0121.000 TRAFFIC MANAGEMENT SPECIFICATION, Version 4.1, March 1999

[10.3] ITU Recommendation E.736 METHODS FOR CELL LEVEL TRAFFIC CONTROL IN B-ISDN, May 1997

[11.1] Rathgeb, E.P., Modeling and performance comparison of policing mechanisms for ATM networks, *IEEE JSAC*, **9**, 3, April 1991, pp. 325–334

[13.1] Kröner, H., Hébuterne, G., Boyer, P. and Gravey, A., Priority management in ATM switching nodes, *IEEE JSAC*, **9**, 3, April 1991, pp. 418–428

[13.2] Schormans, J.A., Scharf, E.M. and Pitts, J.M., Waiting time probabilities in a statistical multiplexer with priorities, *IEE Proceedings – I*, **140**, 4, August 1993, pp. 301–307

[15.1] Braden, R., Clark, D. and Shenker, S. Integrated services in the internet architecture: an overview, RFC 1633, IETF, June 1994

[15.2] Nichols, K., Blake, S., Baker, F. and Black, D., Definition of the Differentiated Services Field (DS Field) in the IPv4 and IPv6 Headers, RFC 2474, IETF, December 1998

[15.3] Braden, R., Zhang, L., Berson, S., Herzog, S. and Jamin, S., Resource ReSerVation Protocol (RSVP) – Version 1 Functional Specification, RFC 2205, IETF, September 1997

[15.4] Schormans, J.A., Pitts, J.M., Scharf, E.M., Pearmain, A.J. and Phillips, C.I., Buffer overflow probability for multiplexed on-off VoIP sources, *Electronics Letters*, 16 March 2000, **36**, 6

[16.1] Floyd, S. and Jacobson, V., Random early detection gateways for congestion avoidance, *IEEE/ACM Transactions on Networking*, **1**, 4, August 1993, pp. 397–413.

[16.2] Schormans, J.A., and Pitts, J.M., Overflow probability in shared cell switched buffers, *IEEE Communications Letters*, May 2000

[17.1] Leland, W.E., Taqqu, M.S., Willinger, W. and Wilson, D., On the self-similar nature of Ethernet traffic (extended version), *IEEE/ACM Transactions on Networking*, **2**, 1, February 1994, pp. 1–15

[17.2] Norros, I., On the use of fractional Brownian motion in the theory of connectionless networks, *IEEE JSAC*, **13**, 6, August 1995, pp. 953–962

[17.3] Mondragon R.J., Pitts J.M. and Arrowsmith D.K., Chaotic intermittency–sawtooth map model of aggregate self-similar traffic streams, *Electronics Letters*, **36**, 2, January 2000, pp. 184–186

Index

Asynchronous Transfer Mode (ATM)
  standards 58
  technology of 3, 8
  time slotted nature 8
  Switches 11

Bernoulli process
  with batch arrivals 19, 87, 98
binomial distribution 16, 89
  distribution of cell arrivals 16
  formula for 16
blocking
  connection blocking 47
buffers
  balance equations for buffering 99
  delays in 110
  at output 99
  finite 104
  sharing and partitioning 273, 279
    see also priorities
  virtual buffers in IP 273
burst scale queueing 125
  by simulation 77
  in combination with cell scale 187
  using large buffers 198

call
  arrival rate 16, 47
  average holding time 47
capacity 49
cell 7
  loss priority bit 205
  loss probability 104, 236, 245
    see also queue
  switching 97
cell scale queueing 113, 121

  by simulation 73
  in combination with burst scale 187
channel 5
circuits 5
  circuit switching 3
connection admission control (CAC) 149
  a practical scheme 159
  CAC in the ITU standards 165
  equivalent cell rate and linear CAC 160
  using 2 level CAC 160
  via burst scale analysis 39, 157, 161
  via cell scale analysis 37, 152, 153
    using M/D/1 152
    using N.D/D/1 153
connectionless service 10
constant bitrate sources (CBR) 113, 125, 150
  multiplex of 113
cross connect 9

deterministic bitrate transfer capability (DBR)
  150
DIFFSERV 12, 253
Differentiated performance 32
Decay rate analysis 234
  as an approximation for buffer overflow
    236
  used in Dimensioning 42, 187
    of buffers 192
effective bandwidth – *see* CAC, equivalent cell
  rate
equivalent bandwidth – *see* CAC, equivalent
  cell rate

Erlang
  Erlang's lost call formula 42, 52
  traffic table 54

Excess Rate analysis (*see also* GAPP)   22, 27
   for multiplex of ON/OFF sources   27
   for VoIP   261
   for RED   271
exponential distribution
   formula for   16
   of inter-arrival times   16

Geometrically Approximated Poisson Process
   (GAPP)   22, 23, 240
   GAPP approximation for M/D/1 queue
     239
   GAPP approximation for RED in IP   271
   GAPP approximation for buffer sharing
     279
geometric distribution   86
   formula for   86
   of inter-arrival times   86
   of number of cell arrivals / empty slots   87
   with batch arrivals   87

Internet Protocol (IP)   10, 58
   IP queueing *see* Packet Queueing
   IP source models, *see* sources models for IP
   IP best effort traffic, ER queueing analysis
     245
   IP packet flow aggregation   254, 255
   IP buffer management   267
   IP virtual buffers   272
   IP buffer partitioning   275
   IP buffer sharing   279
   INTSERV   12, 253

label multiplexing   *see* packet switching
load   47
Long Range Dependency in Traffic   287
   in queue model   293
mesh networks   45

MPLS   12
multiplexors

octets   5

packet queueing   229
   for variable length packets   247
packet switching   5, 229
Pareto distribution   17, 289
   in queueing model   30, 293
performance evaluation   57

by analysis   58
by measurement   57
by simulation   57

Poisson   16
   distribution of number of cell arrivals per
     slot   16, 86
   distribution of traffic   51, 86
   formula for distribution   16
position multiplexing   4
priorities   32, 205
   space priority and selective discarding
     205
     partial buffer sharing   207
     via M/D/1 analysis   207
     push out mechanism   206
   precedence queueing in IP   273
   time priority   35, 218
     distribution of waiting times   2/22, 220
     via mean value analysis   219
     via M/D/1 analysis   220

QoS   253
queue
   cell delay variation   68
   cell delay in ATM switch   68, 108
   cell loss probability in   104
     via simulation   73
   cell waiting time probability   108, 220
   customers of   58
   deterministic   *see* M/D/1
   discard mechanisms   32
   fluid-flow queueing model   129
     continuous analysis   129
     discrete analysis   131
   M/D/1   21, 22, 66, 117
     delay in   108
     heavy traffic approximation   117
     mean delay in   66
   M/M/1   19, 62
     mean delay in   61
     mean number in   62
     system size distribution   63
     summary of formulae   18
   N.D/D/1   21, 115
     heavy traffic approximation   117
queueing
   burst scale queueing behaviour   127
   queue state probabilities   98
   queueing system, instability of   100
   steady state probabilities, meaning of   99

per VC queueing   11, 129
theory, summary   18, 58
Kendall's notation for   60

rate envelope multiplexing   191
rate sharing statistical multiplexing   191
Random Early Discard (RED)   13, 35, 267
Resource Reservation   253
Routers   9, 229

Self-similar traffic   287
simulation   59
    accelerated simulation   77
    cell rate simulation   77
        speedup obtained   80
    confidence intervals   76
    discrete time simulation   59
    discrete event simulation   59
    generation of random numbers   71
        via the Wichmann-Hill algorithm   72
    method of batch means   75
    of the M/D/1 queue   73
    steady state simulations   74
    validation of simulation model   77
sources (also called Traffic Models)   81
source models   16, 81
    as time between arrivals   83
    as rates of flow   89
    by counting arrivals   86
    generally modulated deterministic
        processes   93
    for IP   81
    memoryless property of inter-arrival times
        86
    multiplexed ON/OFF sources   139, 254
        bufferless analysis of   141
    burst scale delay model   145
    ON/OFF source   90
star networks   45
switches (also called Cell Switches)   7, 97

output buffered   97
statistical bitrate transfer capability (SBR)
    10/2
sustainable cell rate   150
Synchronous Digital Hierarchy (SDH)   5

teletraffic engineering   45
time division multiplexing   3
timeslot   3
traffic
    busy hour   51
    carried   47
    conditioning of aggregate flows in IP   265
    intensity   47, 50
    levels of behaviour   81
    lost   47
    models, see sources
    offered   47
traffic shaping   182
traffic contract   150

usage parameter control (UPC)   13, 167
    by controlling mean cell rate   168
    by controlling the peak cell rate   173
    by dual leaky buckets (leaky 'cup and
        saucer')   40, 182
    by leaky bucket   172
    by window method   172
    Tolerance problem   176
    worst case cell streams   178

variable bitrate services (VBR)   150
virtual channels   9, 205
virtual channel connection   9, 205
virtual paths   9, 205
Voice over IP (VoIP)   239
    basic queueing model   239
    advanced queueing model   261

Weighted Fair Queueing (WFQ)   274